SPRINGER PROTOCOLS HANDBOOKS

For further volumes:
http://www.springer.com/series/8623

Principles and Practices of DNA Analysis: A Laboratory Manual for Forensic DNA Typing

Hirak Ranjan Dash

DNA Fingerprinting Unit, State Forensic Science Laboratory, Sagar, Madhya Pradesh, India

Pankaj Shrivastava

DNA Fingerprinting Unit, State Forensic Science Laboratory, Sagar, Madhya Pradesh, India

Surajit Das

Department of Life Science, National Institute of Technology, Rourkela, Odisha, India

Humana Press

Hirak Ranjan Dash
DNA Fingerprinting Unit
State Forensic Science Laboratory
Sagar, Madhya Pradesh, India

Pankaj Shrivastava
DNA Fingerprinting Unit
State Forensic Science Laboratory
Sagar, Madhya Pradesh, India

Surajit Das
Department of Life Science
National Institute of Technology
Rourkela, Odisha, India

ISSN 1949-2448 ISSN 1949-2456 (electronic)
Springer Protocols Handbooks
ISBN 978-1-0716-0273-7 ISBN 978-1-0716-0274-4 (eBook)
https://doi.org/10.1007/978-1-0716-0274-4

This Humana imprint is published by the registered company Springer Science+Business Media, LLC, part of Springer Nature.
The registered company address is: 1 New York Plaza, New York, NY 10004, U.S.A.

Preface

DNA fingerprinting is one of the greatest discoveries of science in the century. Since its discovery, the technique is being widely used in varied fields such as deciphering the genetics of an organism, cracking ancestral belongingness, neonatal diagnosis, diagnosis of genetic disorder, and many others. However, it has revolutionized the field of criminal justice system by its uniqueness and individualization capabilities. Currently, DNA fingerprinting is considered to be the most irrefutable evidence to be produced before the court of law.

Since its inception, the technique has undergone tremendous advancements, and the progress is being observed continuously. However, with the advent of time a consensus step-by-step protocol book describing the technology currently used is lacking. Most of the books available in the same field either describe the obsolete technique of RFLP for DNA examination or manual DNA extraction procedures. Currently, the technological advancement has seen DNA extraction through automation, and RAPID DNA technology has also arrived. Hence, we felt the need of a timely updated laboratory manual in the current field.

In this context, the laboratory manual is structured to clear all doubts of the examiners as well as students regarding conventional and advanced forensic DNA typing experiments. It will be handy for the beginners as well as for the experts in the field of DNA fingerprinting for smooth conduction of the experiments, interpretation, and analysis of results. The uniqueness of the manual involves (a) step-by-step protocol for each experiments; (b) description of each chemicals and reagents and their corresponding principal/functional role; (c) separate experiments for DNA extraction from varied types of biological samples of forensic interest; (d) description of advanced automatic and semiautomatic techniques for DNA extraction; (e) inclusion of real case studies and statistical analysis; and (f) precaution and troubleshooting for each experiment.

Every possible effort has been made to present the protocol book in a very simpler form for easy understanding of students, scientists, and DNA examiners. We have tried our best to incorporate all our experience and expertise to bring out the form of this manual. Throughout the writing process, we have faced lots of hurdles and problems; however, all of them have been overcome due to God's grace, self-belief, and support from family and friends. We are really thankful to each and every one for their support and encouragement during the process of writing. We hope the manual will be of great use for the readers in their career.

Wishing the readers all the best for a successful experiment!!!!

Sagar, Madhya Pradesh, India — **Hirak Ranjan Dash**
Sagar, Madhya Pradesh, India — **Pankaj Shrivastava**
Rourkela, Odisha, India — **Surajit Das**

Contents

About the Authors

Hirak Ranjan Dash is currently a Scientific Officer at the DNA Fingerprinting Unit, Forensic Science Laboratory, in Sagar, Madhya Pradesh, India. He completed his Ph.D. at the National Institute of Technology, Rourkela, India, in 2015, and has since been working in the area of microbial ecology, forensic microbiology, and human genetics. In addition, he is a member of various scientific societies, including the International Society for Forensic Genetics, Association of Microbiologists of India and Students Association of Microbiology. He has been engaged in forensic DNA fingerprinting examinations in criminal and civil cases including the cases related to paternity, child swapping, rape, identity and incest. He has authored more than twenty-five papers in reputed peer-reviewed international journals and has published four books.

Pankaj Shrivastava has broad interests in several areas of forensic DNA fingerprinting including population-based DNA analysis, developing protocols in forensic DNA typing, validating new multiplex kits and optimizing different steps in forensic DNA typing methodology. He has supervised Ph.D. students and published over 40 research papers in peer-reviewed international journals. His publications in Forensic Science International Genetics, Nature Scientific Reports, Legal Medicine, International Journal of Legal Medicine, Annals in human Biology, etc., have earned him wide acclaim from the forensic DNA community. He is presently working at the DNA Fingerprinting Unit in State Forensic Science Laboratory, Govt. of Madhya Pradesh, India. He is a member of International Society for Forensic Genetics (ISFG), Indian Society of Human Genetics (ISHG), Association of Microbiologists of India(AMI) and Association for the promotion of DNA Fingerprinting and other DNATechnologies (ADNAT).

Surajit Das is an Associate Professor at the Department of Life Science at the National Institute of Technology, Rourkela, India. His field of research includes microbial genomics, biosynthesized nanoscale materials, marine biotechnology, environmental microbiology and bioremediation. He is serving as the Editor of Plos One and BMC Microbiology journal. He has published more than 70 research papers in peer-reviewed national and international journals and authored 7 books.

Part I

Overview of Forensic DNA Analysis

Chapter 1

Introduction to Forensic DNA Analysis

DNA fingerprinting analysis has been proved to be a useful tool for crime scene investigation to aid the criminal justice system. With wide media coverage and many high-profile cases solved by DNA technology, it has become familiar among the common people and the awareness to this technology is increasing day by day. The technology not only is helpful in increasing the rate of conviction but also plays an important role in exonerating the innocence.

Deoxyribonucleic acid (DNA) is the genetic material present in every nucleated cell of the body which contains the complete genetic program of the individual for its storage and transmittance. The major constituent of DNA is the nitrogenous bases called as nucleotides, i.e., adenine (A), guanine (G), cytosine (C), and thymine (T), which are arranged in specific order in long sequences. Nucleotides cross-link with each other in a specific manner, i.e., A=T and G≡C. The nucleotide bases reside on the sugar (deoxyribose) and phosphate backbone and subsequently twist around each other to generate a double-stranded helix. A human genome constitutes around 3 billion nucleotide bases connected in a chain-like structure to form a tightly packed chromosome. The number of chromosomes is specific to an organism, e.g., 46 (humans), 38 (cat), 64 (horse), 24 (rice), 48 (potato), and 40 (mango). In this context, genes are the segments of DNA molecules that are responsible for the structure and function of the cell as well as the whole organism. A typical human cell harbors around 50,000–100,000 genes which are spaced at intervals.

1 Genetic Variation

A person inherits his or her genetic material from his or her parents through the cumulative processes of gametogenesis, oogenesis, and

H. R. Dash et al., *Principles and Practices of DNA Analysis: A Laboratory Manual for Forensic DNA Typing*, Springer Protocols Handbooks, https://doi.org/10.1007/978-1-0716-0274-4_1, © Springer Science+Business Media, LLC, part of Springer Nature 2020

zygote formation. Due to the phenomenon of crossing over and genetic recombination occurring at prophase I of the cell division, the genetic variation occurs among individuals as well as siblings. Half of the diploid set of chromosomes of a child comes from the mother and another half from the father. The possible combinations for any parent to transfer its chromosomes to the child is 2^{23} and when both the parents are considered the number of possible combinations further increases to 2^{46} generating a huge set of genetic variations among individuals. Additionally, other factors including mutation, addition/deletion, and substitution of nucleotides also contribute to the genetic variation. In this regard, no two individuals can have the same genetic makeup except the homozygous twins.

Genes, the short DNA fragments of around few to ten thousand base pairs in length, are the coding regions of DNA which constitute ~5% of the human DNA sequence. These genes produce different proteins through the process of transcription which carry out different physiological functions. The protein-coding region of a gene is called exon and the remaining intervening portion is called as the intron. These nonprotein-coding regions present either within a gene or outside the genes are collectively called as junk DNA which are devoid of any necessary functions. The polymorphic variable markers, i.e., loci that are present in the noncoding regions of the DNA, are mostly targeted for the generation of DNA profile of an individual. As human chromosomes occur in pairs and are homologous in nature, a copy of each gene fragment resides at the same position on each chromosome inherited from the parents. The possibilities of each gene fragment occurring at a locus are termed as alleles which may be homozygous (identical alleles) or heterozygous (different alleles) in nature. In this regard, the alleles present at a genetic locus are called as genotype, while a DNA profile is the cumulative genotype found at a set of genetic loci. The current method of human identity testing by DNA typing relies on the use of short tandem repeats (STRs) or microsatellites or simple sequence repeats (SSRs).

2 Short Tandem Repeat (STR) Markers

STRs are repetitive DNA fragments of around 2–7 base pairs which are found abundantly at around 3% of the total human genome. They are considered to be the most polymorphic and prone to mutations due to slippage during DNA replication. Thus, STR markers are considered to be the most suitable for DNA typing analysis as they possess a high degree of interindividual variation and generation of genotypes with greater resolution. Plenty of STR loci occur on each type of chromosomes, i.e., autosomes, and X or Y chromosomes. The STR markers are of various types, i.e., simple repeats (e.g., TPOX, CSF1PO, D5S818, D13S317, D16S539), simple repeats with non-consensus alleles (e.g., TH01, D18S51, D7S820), compound repeats (e.g., vWA, FGA, D3S1358, D8S1179), or complex repeats (e.g., D21S11). Simple repeat STRs contain several units of nucleotides of identical sequence and length, whereas the compound STRs may comprise two or more simple repeats present adjacently. However, the complex STR repeats contain the blocks of several variable unit lengths. Based on the number of nucleotides, the STRs have been categorized into di-, tri-, tetra-, penta-, or hexa-nucleotide markers, out of which the tetranucleotide repeat STR markers are of high use in forensic DNA typing. Additionally, there exists a uniform nomenclature for the STR markers commonly used for forensic DNA typing. For example, if a genetic marker is a part of a gene, the name of the gene is used for the marker designation, e.g., TH01 (present in 1st intron of tyrosine hydroxylase gene), CSF1P0 (6th intron of proto-oncogene c-fms), FGA (3rd intron of alpha-fibrinogen gene), TPOX (10th intron of thyroid peroxidase gene), or vWA (40th intron of von Willebrand factor gene). Similarly, if a STR marker is present outside the gene region, it is named according to its chromosomal position. For a marker present on an autosomal chromosome, e.g., D5S818, "D" refers to DNA, "5" is the chromosome number harboring the marker, "S" stands for the single-copy sequence nature of the marker, and "818" is the order of discovery of the marker on the chromosome. Additionally, for a marker on Y chromosome, e.g., DYS391, "D" refers to DNA, "Y" designates the "Y" chromosome harboring the marker, "S" is the single-copy sequence of the marker, and "391" is the order of discovery of the marker on Y chromosome.

There are many advantages of using STRs over the earlier used genetic markers such as variable number tandem repeats (VNTRs). Generation of small product size due to less number of repeat units of nucleotides increases the chance of recovery of genetic information from degraded samples which is relatively common in forensic cases. A combination of STR markers can be used simultaneously in a single set of PCR amplification process, thus increasing the discrimination power of this technology. Additionally, commercially available STR kits and analysis of a uniform set of core STR loci increase the usefulness of this technology for interlaboratory sharing of results and creation of DNA databases.

3 Technological Use in Human STR Typing

Technological advancements of forensic DNA analysis in chronological order have been given in Table 1.1. The sequential events of forensic DNA analysis involve DNA isolation, quantification, multiplex PCR, capillary electrophoresis, and analysis of results which

Table 1.1
The chronological advancements in forensic DNA typing

Year	Milestone discoveries and advancements
1980	First polymorphic RFLP marker discovered by Ray White
1985	Multilocus VNTR probes described by Alec Jeffreys
1985	PCR technique discovered by Kary Mullis
1986	First case solved in the USA by DNA typing through RFLP technique (Florida vs. Tommy Lee Andrews)
1988	DNA case work lunched by Federal Bureau of Investigation (FBI)
1989	Establishment of Technical Working Group on DNA Analysis Methods (TWGDAM)
1991	First report on short tandem repeats (STRs)
1993	Availability of commercial STR kits
1995	Initiation of DNA database Forensic Science Service (FSS), UK
1996	First use of mitochondrial DNA analysis
1998	Launching of Combined DNA Index System (CODIS) database by FBI
1998	Establishment of Quality Assurance Standards for forensic DNA testing laboratories
1999	Validation of multiplex STRs
2003	Successful application of forensic DNA technology in solving identity crisis in Saddam Hussein case and September 11, 2001, World Trade Center victims
2003	Human Genome Project completed
2003–04	Y STR kits launched
2007	Launching of first kit to analyze samples with degraded or limited DNA
2010	X-STR kit launched
2012	Development of forensic test to predict the hair and eye color of the suspect(s) based on the DNA trace left at the crime scene
2013	Introduction of the portable system to analyze human DNA within 75 min
2014	Developmental validation on autosomal STR multiplex kits with Y STRs and InDel markers
2016	Launch of NGS-based STR kits
2017	Expanded CODIS for inclusion of 20 STR loci

has been discussed in much details in the subsequent chapters. In a nutshell, the preliminary requirement of forensic DNA analysis is the isolation and purification of DNA from the forensic case exhibits. In any cell, DNA occurs in the form of either nuclear DNA or extranuclear DNA (e.g., mitochondrial DNA or chloroplast DNA of plants). Though various DNA isolation techniques are available nowadays the procedure utilized is exclusively dependent on sample, technique, and analyst's preference. DNA isolation can be carried out by either manual methods using different sets of chemicals and reagents or automatic or semiautomatic techniques depending on the nature of the sample. Isolation of DNA is followed by both qualitative and quantitative assessment of the extracted DNA. Input of optimal quality and quantity of DNA into the PCR reaction is the key to generate an acceptable DNA profile. Hence, various quantitation techniques such as UV absorbance, gel electrophoresis, quantitative real-time PCR, and others are being used nowadays. Many qRT-PCR-based kits are also commercially available for the simultaneous detection of many useful parameters such as amount of total human DNA, ratio of amount of small-to-large autosomal DNA reflecting the degradation index of the DNA sample, male-to-female DNA ratio in mixed samples, and presence of PCR inhibitors in the sample.

Quantification of DNA is followed by the amplification of target loci by polymerase chain reaction (PCR). In this process, simultaneous amplification of multiple DNA fragments is carried out using different dye-labeled primers, thus generating PCR products labeled with dyes. Many multiplex kits are available currently for amplification of desired markers either on autosomal, X, or Y chromosome. Subsequently, the PCR products are subjected to capillary electrophoresis for separation and detection of the amplified STRs. In this process, DNA fragments are separated based on their molecular size under voltage and subsequently size of the amplified fragments is designated by comparing them with the standards. Finally, the results obtained were analyzed using various analysis software to generate the desired results. The flowchart describing the sequential events in forensic DNA typing has been given in Fig. 1.1.

4 Challenges in Forensic DNA Typing

Ideal samples are rarely received in a forensic DNA laboratory for examination except the reference samples. Most of the samples contain either degraded DNA or potential inhibitors of PCR process. In certain instances, the samples are highly contaminated from exogenous DNA and pose a great challenge to the process. Other major problems of forensic DNA typing include the generation and analysis of mixed DNA profiles, cost-effectiveness, and requisite expertise. A few of the most important challenges related to forensic DNA typing have been discussed below.

Forensic casework encounters the major chance of examining degraded biological samples. Primarily, two factors, i.e., time and condition of exposed environment, influence on the decomposition as well as degradation of DNA evidences. The process of degradation increases with time, whereas the environmental conditions such as temperature, humidity, and pH influence the rate of

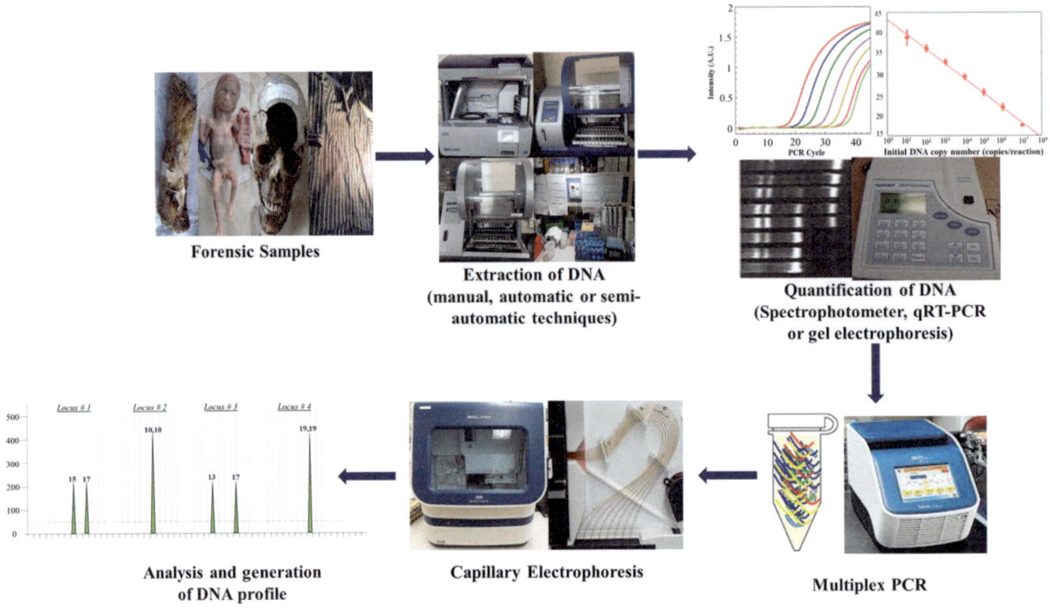

Fig. 1.1 The generalized steps involved in forensic DNA typing

degradation of the samples. In cases related to burning and mass disaster generally degraded samples are received for examination. As the biological samples degrade, the DNA becomes more fragmented and the chance of generating a complete DNA profile is minimized. Most of the common forms of DNA damage include oxidation, hydrolysis, pyrimidine dimers, cytosine deamination, and formation of cross-linkages between DNA-DNA and DNA-protein. In forensic samples DNA damage occurs mostly due to oxidation, hydrolysis, and pyrimidine dimer formation.

In certain instances, despite the presence of adequate amount of DNA, the amplification failure takes place leading to non-generation of a DNA profile. Presence of PCR inhibitors co-extracted with the target DNA commonly include the dyes from clothing, humic acid from soil, fatty acid from bone, heme from blood, and melanin from hair. Presence of inhibitors in a sample may lead to dropout of alleles at some loci. Other problems associated with the inhibitors include adverse effect on PCR primers, overlysis of cells during DNA extraction, and/or reduction of polymerase activity. Some of the common strategies to overcome PCR inhibitors include dilution of the extracted DNA to reduce their content, use of robust, inhibitor tolerant Taq polymerase, use of sample cleanup devices, as well as application of chemicals with PCR-enhancing properties such as bovine serum albumin and PCRBoost™. Magnetic bead-based DNA extraction technique can be used as it possesses the property of highest inhibitor removal capability.

Forensic DNA typing is prone to low levels of contamination. Contamination during forensic DNA typing originates from either primary or secondary sources. In certain instances, contaminants in the form of either cells or purified or amplified DNA get accumulated on various equipment leading to their transfer to the questioned samples. The exogenous DNA can be categorized into four types, i.e., (1) DNA originated from the examiner, (2) DNA originated from consumables (pipettes, plasticwares or chemicals) being used (3) DNA from other samples analyzed together, and (4) DNA from mixing of allelic ladder with the questioned samples. Contamination plays a limiting role in the analysis and interpretation of DNA typing results. The presence of unexpected peaks, occurrence of additional peaks in a single source sample, and heterozygous imbalance are the most common problems of exogenous DNA contamination. A little bit of precaution and the use of certain chemicals may minimize the risk of contamination during forensic DNA analysis. Thorough cleaning of working areas, decontamination or autoclaving of the equipment and instruments, and application of chemically decontaminating solutions such as sodium hypochlorite (NaOCl) can reduce the chance of contamination from exogenous sources. Additionally, the complete procedure of DNA typing from DNA isolation to result interpretation involves a huge cost, thus minimizing its usability in developing nations. Technological complications lead to the examiners to gain expertise in a very slow pace. Cost involvement in infrastructure development also creates a hindrance to adoption of DNA typing technology in many cases, mostly in developing and undeveloped countries.

5 Applications of Forensic DNA Typing

DNA fingerprinting has revolutionized the process of justice delivery system. DNA typing is a reliable technique to identify the criminal suspect(s) and possesses the potential to either implicate or exonerate a suspect with greater degree of fidelity. The technology can be applied to deduce conclusion in a variety of case types including paternity dispute, identification of mutilated bodies, murder, abduction, and sexual assault. Hence, currently DNA typing is routinely used in most of the civil as well as criminal cases. Some of the important high-profile cases solved through DNA typing have been discussed below.

5.1 O.J. Simpson Case in Los Angeles

On 12th June 1994, Nicole Brown Simpson and Ronald Goldman were found murdered in the house of Ms. Simpson. During investigation, O.J. Simpson, the ex-husband of Ms. Simpson, was found to be the prime suspect of the incidence. As O.J. Simpson was a successful footballer and celebrity of that time, the case drew attention of the global public. Many samples including the blood droplets and the stains were collected from the spot for examination. The samples were examined in three different laboratories, i.e., Los Angeles Police Department (LAPD), the California Department of Justice (CA DOJ), and a private laboratory for forensic DNA typing. Both PCR- and RFLP-based techniques were employed. Finally, a consensus result was obtained from the three laboratories with no exclusion between the samples collected from the scene of crime and reference sample collected from Mr. Simpson. A long hearing courtesy to the advocates of Mr. Simpson took place in this case on raising the contamination theory for collection of scientific evidence. However, a random match and obtaining of such overwhelmingly incriminating results are not due to mere contamination and random laboratory error. This case is popularly known as the case of the century and was a major boost for advocating implementation of DNA typing in criminal justice system.

5.2 Clinton-Lewinsky Affair

This was an American political sex scandal case which involved the then president of America, Bill Clinton, and an intern Monica Lewinsky working at the residential office of the president. The sexual relationship that occurred during 1995–1997 came to public in 1998 and Linda Tripp, a co-worker of Monica, was familiar with the situation. When president denied the allegation in public and Monica almost submitted an affidavit refuting a relationship with Clinton, Linda made this case public. Earlier, Linda had encouraged Monica to save the semen-stained blue gown without dry cleaning which later turned out to be the most valuable scientific evidence. Subsequently, a DNA test was carried out and the DNA profile obtained from seminal stain present on the blue gown matched with the reference sample obtained from the president.

5.3 Identification of the Bodies in World Trade Center Attack

The mass catastrophe occurred after attack on the World Trade Center on 11th September 2001 resulted in the generation of many mutilated charred bodies posing a huge challenge for their identification. There was no definite list of victims. Hence, an attempt was made to process as many samples recovered separately assuming them to have a distinguished origin. At the early stage, the soft tissues were targeted for identification purpose. Cross contamination of the samples due to the high explosion in the spot yielded a mixed STR DNA profile which shifted the target of identification to bone or skeletal remains rather than the soft tissues. In this regard, the intact bones were preferred over the spongy bones. Out of the samples analyzed the rate of success in generating a DNA profile was 75%, leading to the identification of the charred remains.

5.4 Assassination of Sri Rajeev Gandhi, Prime Minister of India

India's former Prime Minister Mr. Rajiv Gandhi was assassinated in a public meeting at Sriperumbudur, Tamil Nadu, India, in May 1991 which shocked the nation. This is one of the earliest cases of Indian history to be solved by DNA fingerprinting technique. The mastermind of this case was Sivarasan, the "one-eyed jack" from Sri Lanka. After 3 years of the incidence, Sivarasan killed himself along with his associates. The identification of Sivarasan was established by comparing his DNA profile with that of his mother Sivapakkam and brother Ravichandran. This settled the identification issue of Sivarasan and gave a breakthrough to the investigating agency. Another important aspect of this case was that a girl named Dhanu was used as the human bomb for this assassination purpose. However, post-incidence the investigating agencies were clueless about certain questions, e.g., whether various pieces of the body that were put together during investigation belong to one individual, whether the denim belt which contained charred muscle pieces was worn by that individual, and whether it is a male or female individual. All these questions were answered with great level of satisfaction through DNA fingerprinting technique which established Dhanu as the suicide bomber in this case.

Chapter 2

Biological Samples: The Target Sources for DNA Typing

Human body contains around 10^{13} number of cells. All nucleated cells of the human body contain DNA and hence any biological material left in the crime scene can be collected and used as a biological evidence for DNA typing analysis. However, the samples should be collected properly to avoid contamination or degradation. With the advent of PCR technology, a trace amount of biological evidence present in a crime scene can also be useful due to the amplification process and reaching the detection limit of the instruments. In this regard, touch DNA technology is also in practice currently for small amount of DNA present in the samples.

1 The Locard's Principle

In the twentieth century, a pioneer forensic scientist Dr. Edmond Locard formulated the theory of trace evidence which states that "every contact leaves a trace." These fragmentary evidences can be used to link between the victim, suspect(s), spot, and the weapon (if any). In the process of committing a crime, the perpetrator may leave some biological samples of his/her own in the crime scene either knowingly or unknowingly. Thus, a proper investigation and search in the spot as well as the victim's belongings may generate a clue in the form of biological evidence to link the suspect(s). Some of the biological evidences in the form of either primary sources or secondary sources have been discussed in details subsequently.

2 Types of Biological Samples

Any object or its surface containing nucleated cells can be used a source sample for DNA typing analysis. These exhibits can be

H. R. Dash et al., *Principles and Practices of DNA Analysis: A Laboratory Manual for Forensic DNA Typing*, Springer Protocols Handbooks, https://doi.org/10.1007/978-1-0716-0274-4_2, © Springer Science+Business Media, LLC, part of Springer Nature 2020

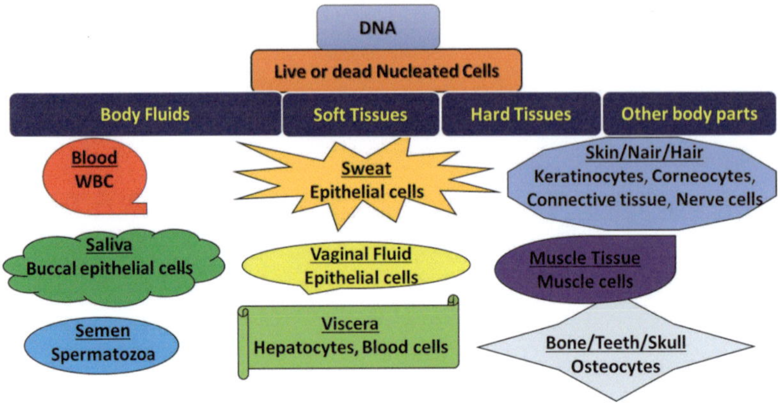

Fig. 2.1 Samples and their source or DNA for forensic analysis

collected either from the crime scene or from the living human beings with their consent. Biological evidences of forensic use are of varied nature, i.e., minute objects such as hairs and nails to large objects such as skeletons and clothing. The physical nature of these biological evidences may be either solid or liquid. The biological samples can be collected instantly under proper conditions. On the contrary, some biological evidences are collected after a long period of exposure in the environment. Though the core chemical composition of every cell of human body is the same, the amount of the constituents varies according to the requirement of the body parts. But the vital molecule DNA is present in equal amount in every body cell except the gametes. The physical and chemical composition of few biological source samples has been given in Fig. 2.1.

Additionally, the full forensic value of the evidence is achieved only when it is compared with the reference DNA profile. In this regard, the reference samples are also to be collected in the form of either whole blood, buccal swab, or hair. An extensive list of physical evidences that can be collected as a source of DNA sample has been provided in Table 2.1.

3 Whole Blood/Blood Stains

Blood is a specialized fluid connective tissue, constituting 7% of adult body weight which travels through the circulatory system transporting gases, nutrients, wastes, and other macromolecules throughout the body. Human blood has a solid phase of formed elements and a liquid phase, the plasma. Plasma, the straw colored aqueous solution, is mostly made up of water (90%), proteins (7–8%), and other solutes (1–2%). Blood cells (formed elements) are classified as either erythrocytes (RBC) or leukocytes (WBC). Additionally, a smaller third cellular constituent is also present referred to as thrombocytes (platelets). There occur approximately

Table 2.1
Sources of biological evidences with potential use in forensic DNA typing

Physical evidences	Location within the object	Source of DNA
Consumed cigarette	Cigarette butt	Saliva
Tooth pick	Tips	Saliva
Stamps and envelopes	Licked area	Saliva
Bottle or sipper	Mouthpiece	Saliva
Bite mark	Person's skin	Saliva
Used condom	Both the sides	Semen and vaginal cells
Blanket/bedsheets	Surface	Semen, sweat, hair, saliva, urine
Fingernail	Scrapings	Blood, skin
Ligature	Surface	Skin
Bullet	Surface	Blood
Clothing	Surface	Blood, sweat, semen
Hat/mask	Inner surface	Sweat, hair
Knife	Outside surface/handle	Blood, tissue, sweat
Credit/debit card	Surface	Sweat, tissue
Gloves	Interior and exterior	Sweat, tissue
Toothbrush	Surface	Buccal tissue, saliva
Dandruff	Whole	Cell debris
Head comb	Surface	Sweat, hair, tissue
Razor	Blades	Tissue, hair roots
Eye glasses	Ear and nose piece	Sweat, tissue

4.5–5 million RBCs, 4500–11,500 WBCs, and 250,000–450,000 platelets per microliter of blood. WBCs, as a part of body's immune system, destroy and/or remove cellular debris and infectious foreign pathogens. The source of DNA in blood is only the leukocytes, i.e., WBC.

Bloodstains can be obtained either from the spot, on the weapons, or from the clothing. From forensic point of view, the shape and pattern of bloodstains give useful information regarding the origin and nature of the wound occurred during crime. Additional information regarding the distance of blood fall, bleeding person's position, i.e., standing, still, or walking, etc. can also be deduced from careful observation of the bloodstains. Bloodstains are of two types, i.e., wet bloodstains and dry bloodstains. However, both of them act as a source of DNA for forensic DNA typing.

4 Hairs

Hairs grow from the follicles found in the dermis which is the exclusive feature of the mammals. The primary composition of human hair is a protein, i.e., alpha-keratin. A typical hair has two structures: the follicle or bulb which is present beneath the dermis and the shaft which is the hard filamentous structure that is present above the sin surface. Several layers exist in a hair fiber which includes the cuticle, cortex, and medulla. Though there exist many forms of hair in the human body, scalp and pubic hairs are the routine candidates for DNA examination.

Hairs are considered to be the most overestimated and misinterpreted samples for DNA examination. Many investigators assume hair to be an ideal sample for DNA examination. However, the extraction of DNA and subsequent analysis depend primarily on the part of DNA that has been received for examination. Isolation and characterization of nuclear DNA are possible only in the hairs with growing roots and adhering tissues. However, forensic mitochondrial DNA analysis relies on the DNA content of the hair shaft rather than hair root. During the process of keratinization, nuclear cellular material is degraded in the hair shaft and the presence of multiple numbers of mitochondria makes them a great source of mitochondrial DNA for further analysis. In this regard, it is highly recommended for microscopic examination of hairs prior to their DNA analysis regarding their suitability.

5 Teeth

With the advancements in DNA technology, tooth becomes one of the most useful sources of DNA for its subsequent analysis. Human teeth are considered to be the most preferred source of DNA due to their unique composition and location within the jawbones which prevents the decaying process from the physical and chemical agents. The advantages of the DNA extracted from teeth are their high quality and being less prone to contamination.

Human teeth have two parts, i.e., the crown which is exposed in the mouth and the root present in the jawbones. Teeth roots are primarily composed of cementum and pulp which are a rich source of DNA. It has been documented that the DNA yield from the teeth root is ten times higher than that of the DNA yield from the crown. Enamel, the hardest tissue of human body, covers the crown of the teeth and is acellular in nature, and hence contains no DNA. The tissues that primarily occur in the teeth pulp are the odontoblasts. Other cells that occur in traces in human teeth include the fibroblasts, histiocytes and macrophages, plasma cells, nerve cells, and undifferentiated mesenchymal cells.

6 Bone

Bones are the sources of DNA in case of the extremely damaged or degraded human skeletal remains. Many exogenous elements such as humidity, organic compounds, and microorganisms reduce the amount of informative DNA on the soft tissues and other parts of corpse, thus making bones the most suitable samples for genetic analysis. Bone possesses a complex anatomical structure. The basic building block of bone includes the collagen fibrils. Additionally, vascularization also contributes to the microstructure of bones. Bone constitutes many organic as well as inorganic molecules such as hydroxyapatite $[Ca_{10}(PO_4CO_3)_6(OH)_2]$, collagen, and non-collagen proteins such as osteonectin and osteocalcin. Often DNA gets adsorbed to the inorganic constituents like hydroxyapatite for its long-term storage by developing certain crystal structures. Additionally, DNA binding to the collagens increases the stability of the DNA. Certain short fragments of DNA get trapped in the bone fibrils by encapsulation, increasing its stability. Ultimately, the combination of organic as well as inorganic substances renders bones their hardness and makes them less prone to degradation after extensive environmental exposure. In this regard, bone samples provide a useful source of DNA in many forensic cases such as identification of deceased individuals in cases of accidents, mass disasters, terrorist attacks, and cases involving defragmentation of the body.

7 Soft Tissues

Though hard tissues of human corpse such as bone and teeth are considered to be the most suitable samples for DNA typing studies, soft tissues are also highly useful in many instances. The usefulness lies in the fact that they contain a huge amount of DNA and it takes less time to isolate DNA from them in comparison to bone samples. Soft tissues can be collected from the fresh corpses with minor decomposition. Soft tissues are also preferred for the identification of unknown fetus aborted illegally. The most advised soft tissues for DNA analysis include the brain, muscle (biceps), kidney (from upper pole), and heart (right ventricle). The tissue samples are most suitable for DNA isolation due to their syncytial nature which yields a higher DNA content in comparison to other body parts. In this regard, the viscera collected in cases of unnatural death to determine the cause of death are also useful samples for DNA tests. However, the base modification and the presence of reactive oxygen species (ROS) in the viscera samples generated due to the toxic effects may hinder the DNA technology and further downstream processing of the samples.

8 Semen and Sperm/Vaginal Fluid

Most of the forensic DNA laboratories receive nearly half of their cases as the sexual assault cases. In such cases, the mixture of seminal and vaginal fluids is the most likely sample to be received for DNA testing in the form of either slide, swab, or clothing. Sperms are considered to be the most useful source of DNA in a very small quantity. A very small amount of seminal stain can yield a good DNA profile as millions of sperm cells are present per microliter of human ejaculate. The time elapsed during the coitus and collection of sample is of utmost importance as semen can last in the victim's body for 72 h or more. In addition to the spermatozoa many epithelial cells are also expected from the sexual assault cases. Human seminal fluid contains acid phosphatase, citric acid, inositol, calcium, zinc, magnesium, fructose, ascorbic acid, prostaglandins, and other compounds. Additionally, vaginal fluid is a mixture of cervical mucus such as albumin, alpha1-antitrypsin, alpha2-haptoglobin, alpha2-macroglobulin, beta-lipoprotein, orosomucoid, and ceruloplasmin. The epithelial cells of both the male and female contributors also provide the source of DNA in these samples.

9 Urine

Urine samples are the most easy to collect from a person. Though urine samples are used for analysis of biomarkers for various types of diseases, they have limited use in forensic DNA applications. Urine is not being considered for forensic DNA applications due to the low content of nucleated cells in them. The major constituent of human urine is water, urea, chloride, sodium, potassium, creatinine, and other organic and inorganic compounds. Most of the scanty nucleated cells found in urine are in the form of either white blood cells or epithelial cells originated from renal tubular, transitional urothelial, and squamous cells. Urine of females consists of a larger amount of epithelial cells in comparison to males. Urine samples may be of useful forensic implications if DNA can be isolated from the urine-stained clothing in cases of either sexual assault or murder.

10 Saliva

Saliva is the biological secretion of mouth helpful in the digestion process. The major constituents of saliva include urea, glucose, progesterone, various traces of acids, amino acids, creatinine, and different proteins. It also contains a high amount of buccal cells that contribute to the source of DNA. Saliva generates a smaller amount of DNA than blood and other body fluids. However, its ease of collection, requirement of simpler logistics, chance of self-collection, and sample mailing make them a suitable candidate for forensic DNA analysis.

11 Biological Samples and Their DNA Content

The advent of PCR technology has revolutionized the DNA analysis by reducing the requirement of initial amount of DNA template to a substantial limit. In this regard, an overview on the approximate DNA content of the biological samples has been given in Table 2.2.

A diploid human cell contains approximately 6 pg of DNA. For blood, the source of nuclear DNA is only WBC. Under normal circumstances, human blood contains 5–10 $\times$ 10^6 WBCs/ml. Hence, theoretically, 1 μl of blood yields 30–60 ng of DNA. However the amount of DNA present in bloodstains varies greatly with storage condition of the stain and stage of degradation of the sample. Similarly, sperm cells, being haploid in nature, contain around 3 pg of DNA. The stains containing the mixture of seminal/vaginal fluid are a great source of DNA for sexual assault cases. Additionally, DNA content in hairs varies greatly between

Table 2.2
Biological samples of forensic implications and their DNA content

Source samples	DNA content (approximately)
Liquid blood	30–60 ng/μl
Blood stain	250–500 ng/cm^2
Semen	150,000–300,000 ng/ml
Vaginal fluid	Up to 3000 ng/ml
Saliva	1000–10,000 ng/ml
Oral swab	1000–1500 ng/ml
Hair root (plucked)	49–72 ng/μl
Hair (shed)	1.72–1.872 ng/μl
Dandruff	0.8–16.6 ng/particle
Urine	1–20 ng/ml
Bone	3–10 ng/mg
Teeth	21–37 μg/ml
Soft tissues	3–15 μg/mg
Fibroblast cell line	6.5 μg/l

individuals as well as different hair types of the same individual. The approximate amount of DNA content of hair samples varies between 49 and 72 ng/μl, whereas storage reduces the DNA content to approximately 1.72–1.87 ng/μl.

The content of total DNA varies greatly among individuals and teeth types. Other factors influencing the DNA content of a tooth sample include the tooth type, age of the individual, as well as health status of the individual. Studies have shown that the teeth with large pulp volume yield a highest amount of DNA. Thus, multirooted teeth such as premolars and molars are the suitable sources to be subjected for DNA test. The number of odontoblasts in teeth pulp tissues is approximately $11,000/mm^2$ and that of fibroblasts is around $1000/mm^2$.

DNA content of bone samples is really scanty. A fresh bone sample may contain 3–10 ng/mg of DNA; however, the presence of high quantity of PCR inhibitors limits their usefulness. With the increase in duration of exposure in the environment, the DNA content of a bone is decreased to a substantial level. Despite all the odds, the usefulness of bone samples for extraction of DNA is evidenced from its application in archaeological and paleontological studies. Human urine contains very trace amount of DNA and the extractable amount of DNA varies greatly with the methodology used. The average amount of extractable DNA varies from 53 to 200 ng/ml for females and from 3 to 50 ng/ml for males. Similarly, the DNA content of saliva in children is of lower amounts than in the adults. Studies have shown the quantity of DNA obtained as 34.91 μg (2.20–122.04 μg) from 3 ml of saliva.

Chapter 3

Collection, Transportation, and Preservation of Biological Evidences for DNA Analysis

Application of DNA technology has revolutionized the field of forensic science. To obtain an appropriate result from any biological evidence, proper collection, transportation, and preservation/storage strategies should be employed for respective samples without manipulating the chain of custody (Fig. 3.1). Many factors contribute to the generation of a good DNA profile from a biological sample, i.e., the sensitivity of a PCR setup, rate and extent of degradation, purity, and DNA content of the sample. No biological evidence is resistant to degradation. Thus, careful collection and proper storage of the biological samples can provide a useful information when subjected to DNA typing analysis. Microbial or environmental degradation does not alter the genetic profile of a subject, but can impact adversely by obtaining a partial DNA profile from the sample. Evidences suitable for DNA examination are limited to biological samples; hence, a proper precaution for collection, transportation, and storage is required for these samples which has been discussed in details in this chapter.

1 Liquid Blood from Donor

In most of the DNA typing experiments blood samples collected from the victim as well as the suspect(s) are considered to be the most suitable samples. Mostly, reference samples are collected in the form of liquid blood. Liquid blood is collected with the consent from the donor by a qualified medical practitioner. Certain medical parameters such as blood or organ transfusion of the donor should be communicated to the laboratory along with the samples. Blood

H. R. Dash et al., *Principles and Practices of DNA Analysis: A Laboratory Manual for Forensic DNA Typing*, Springer Protocols Handbooks, https://doi.org/10.1007/978-1-0716-0274-4_3, © Springer Science+Business Media, LLC, part of Springer Nature 2020

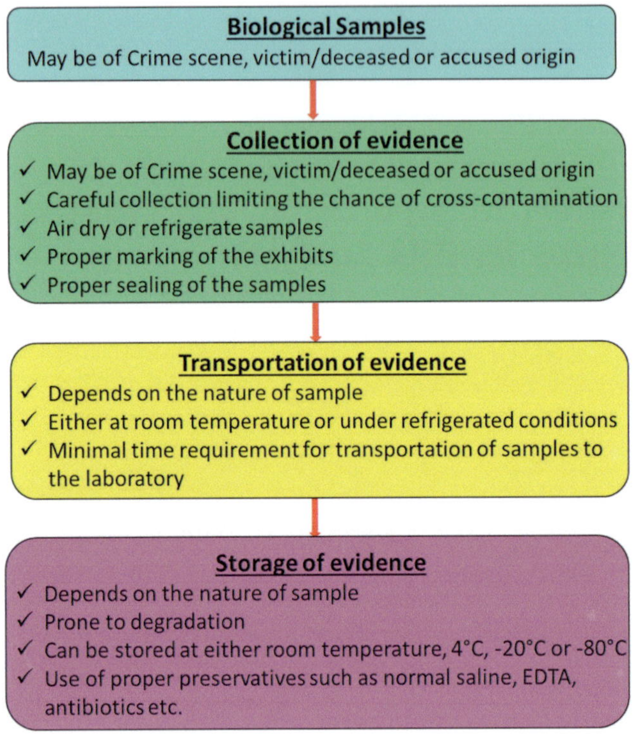

Fig. 3.1 The sequential events unfolded prior to DNA examination of biological samples

sample of around 2 ml should be collected in duplicates in vacutainers with 5 mM EDTA, sodium citrate, and heparin as anticoagulants. Blood sample can also be collected from the postmortem dead body in a similar fashion. In such cases, it should be collected from non-body cavity areas such as heart or major internal blood vessels till 24 h of last heartbeat. After collection, each tube should be labeled with date, time, subject's name, location, collector's name, crime and exhibit number, and sealed properly. It should be immediately refrigerated (not frozen) and reach at the laboratory at the earliest.

Reference samples in the form of liquid blood can also be collected on Whatman FTA® cards with appropriate labeling. Liquid blood can be dropped on the card in a concentric circular motion inside the marked circular area. In general, <125 μl of blood should be placed per inch of the circle on FTA card. It is recommended to dry the card for 3–5 min before placing the cards separately on individual labeled envelopes. These cards can be transported to the laboratory at room temperature. The condition of transportation and storage of biological exhibits has been given in Fig. 3.2.

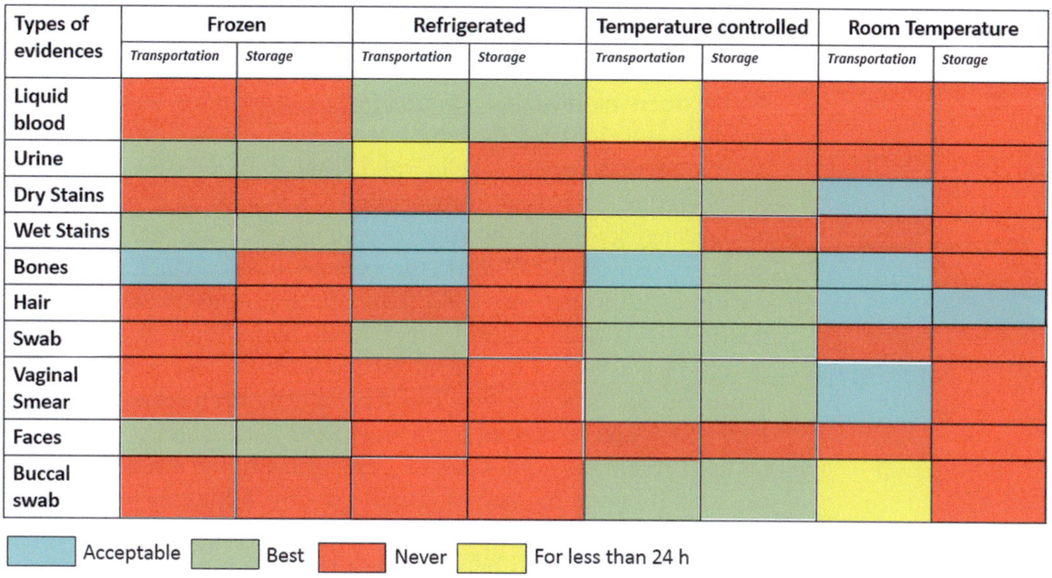

Fig. 3.2 Recommended physiological conditions for transportation and storage of biological evidence for forensic DNA analysis

2 Liquid Blood and Wet Bloodstains from Crime Scene

In certain instances, prompt visit to crime scene may encounter the investigating agency with the presence of liquid blood. Liquid blood can be collected by a sterile syringe and transferred to a clean tube preferably containing the anticoagulants. Similarly, a blood clot can be collected with a sterile spatula in a tube. Alternatively, the liquid blood present around a blood clot can be collected by a clean cotton cloth by soaking without touching the areas containing serum. The collected samples should be refrigerated and sent to the laboratory at the earliest. Additionally, the collected samples should be labeled with crime and exhibit number, date, time and location of sample collection, and name of the collector. Later seal the samples properly prior to sending them for laboratory examination. In many cases, liquid bloodstains can be found on small or large objects at the crime scene, where the bloodstains should be allowed to dry completely followed by their collection without altering their integrity. If the stain is present on large objects, the bloodstain should be collected using a moistened sterile cotton swab. The swab should be air-dried completely before packaging followed by proper labeling and sealing of the object.

3 Dried Bloodstains

If the dried bloodstains are present on removable items such as weapons and clothing, the entire object can be collected on separate containers with proper labeling and sealing. Bloodstains should be considered to have their distinguished origin and hence they should be collected separately to avoid contamination. If the stain is present on solid immovable objects, the pattern of bloodstain should be documented prior to its collection. The stain can be tape lifted or scrapped on a clean paper. In such cases, blood crust and tape lifter should be placed in a druggist fold and sealed in an envelope. If the condition arises where the bloodstains can be neither lifted nor scrapped, a sterile cotton swab moistened with saline or water can be used to elute the bloodstain. After elution, the swab should be air-dried properly to prevent fungal growth. Additionally, a control swab collecting sample from the adjacent area without containing the bloodstains should be sent along with the sample for examination. If the object having the bloodstain can be cut, the portion of the item containing the bloodstain should be cut using a sterile sharp instrument. Additionally, a control piece containing the unstained portion of the object should also be sent for examination. The collected samples should be labeled and sealed properly and can be stored and transported to the laboratory at room temperature.

4 Semen, Seminal Stains, and Evidence from Sexual Assault Victims

Prior to the collection of seminal stains on an object it is always recommended to document them by either photography, videography, or sketching. For the presence of liquid semen, a fresh syringe or disposable pipette can be used to transfer the liquid to a fresh tube. The tube containing the liquid semen should be properly labeled, sealed, and stored under refrigerated condition. In another manner, the liquid semen can be adsorbed on a sterile cotton cloth, air-dried, labeled, and sealed properly. Dried seminal stains present on small movable objects such as clothing, bedsheet, and pillows can be collected as such. If any of the articles contain stains in wet condition, it should be thoroughly air-dried prior to its proper labeling and sealing. If the stain is present on large immovable objects, it should be either cut using sterile scissor or scrapped using a sterile scalpel, and kept at sealed container with proper labeling and sealing.

For sexual assault cases, the medical examination of the victim should be carried out at the earliest. Based upon the condition, requisite samples such as oral, vaginal, and/or anal swabs should be collected, air-dried, packed, labeled, and sealed properly. In a similar way the victim's clothing can also be preserved. The main reason to conduct an early medical examination is that victim's body initiates breaking down of the seminal fluid through drainage, enzyme activity, pH, etc. The purpose of air-drying the objects is to minimize the growth of microorganisms. Other body fluids such as urine, saliva, and others should be collected in a similar fashion as that of the blood and bloodstains.

5 Soft Tissues and Organs

If any tissue or organ sample is collected from the crime scene, it should be picked up in clean, sterile forceps, and kept separately in a glass bottle in pre-sterilized normal saline solution without any other fixatives. Optionally antibiotics (e.g., 0.1 mg/ml gentamicin solution) can also be added to the normal saline solution to prevent microbial growth. Each container should be labeled and sealed properly and stored at room temperature to reach the laboratory at the earliest. If soft tissues and organs are collected from a postmortem subject with advanced decomposition state, a portion of deep muscle tissue with more than one organ should be preserved in the aforementioned procedure. Preferred organ tissues for collection are heart or brain. Liver and kidney are generally not recommended samples for DNA examination. The collected samples should be refrigerated immediately either with normal saline solution or without any additives. It is always recommended not to store the tissue samples in formalin.

6 Teeth and Bone

Teeth and bone samples can be preserved from skeletonized structures after confirming their human origin from the autopsy surgeon. 2–4 intact (rooted) molar teeth are recommended for collection in such cases. Additionally, compact bone (most preferably femur) should be collected for the purpose of conducting DNA test. Bone and teeth samples can be stored and transported at room temperature to the laboratory after proper labeling and sealing.

7 Hair

For examination of hair samples by nuclear DNA analysis, it should contain the root sheath. If hairs are found in the crime scene, it should be collected with precaution in any paper packet to be placed in an envelope. However, it is recommended to submit 15–20 representative hairs to the laboratory as sometimes additional microscopic examination may be required for the same. The most important thing to keep in mind is that the hair samples should be followed by the reference samples (blood/buccal swab) of the suspect(s) for DNA match.

The guidelines and precautions for collection, transportation, and preservation of biological samples for DNA examination have been given in Table 3.1.

Table 3.1
Guidelines and precaution for proper collection, transportation, and preservation of biological samples for forensic DNA examination

Biological samples	Physical state/ condition	Source	Collection and preservation	Transportation	Precaution
Blood	Liquid	Living person	2–5 ml in duplicates in EDTA vial/FTA card	Under refrigerated condition, to reach laboratory at the earliest (24–48 h)	Use of disposable syringe
	Liquid	Autopsy sample	2–5 ml in duplicates in EDTA vial/ FTA card		
	Liquid	Crime scene	In duplicates in EDTA vial/FTA card		
	Wet cloth	Crime scene	Collect in sterile tube, either add equal volume of normal saline solution or air-dry and transport at room temperature		Do not handle clothing with bare hand
	Wet	Clothing of victim and suspect(s)	Air-dry thoroughly at room temperature, pack separately	Must reach the laboratory at the earliest and can be kept at room temperature	Do not use direct sunlight, hot air blower, or heater to dry
	Wet	Object such as knife			
Dried blood and other body fluid stains	Crust/stain/spatters	Crime scene/ unmovable objects	Scrap onto a paper and pack in an envelope; transfer onto a moisten cotton swab, air-dry and pack in an envelope, collect proper negative control from adjacent area		Document stain patterns, do not mix stains from different locations
	Stain	Weapon and on small movable objects	Allow the stains to dry and collect the whole item		
	Stain	Vehicle, carpet, wooden objects	Cut the stained area, air-dry, collect a negative control from the adjacent area		
Semen	Liquid	Victim	Collect with sterile cotton swab, air-dry, pack in envelope		
	Liquid	Crime scene			

Tissue/organ	Wet/semisolid	Dead body/mutilated remains	Tissue samples to be collected in normal saline, refrigerate	Submit to the laboratory under refrigerated condition at the earliest	Do not preserve samples in formalin
Bone/teeth	Solid	Dead body/crime scene	Clean to remove debris, air-dry completely, pack in paper envelope	Must reach the laboratory at the earliest and can be kept at room temperature	Completely charred bones are not suitable for DNA examination
Rooted hair	Dry	Crime scene, weapon, victim/suspect(s) body, and clothing	Collect by sterile forceps and pack in envelope		Cut hairs are not suitable for DNA examination

Source: FSL Bulletin Forensic Science Laboratory, vol. ii, no. iv, June 2008

Part II

Manual Extraction of DNA from Biological Samples

Chapter 4

Manual Isolation of DNA from Whole Blood

Objective: To isolate DNA from whole blood (liquid) by manual isolation method.

1 Introduction

A typical forensic DNA analysis involves the matching of DNA profile between the subject and the reference samples. Reference samples are collected from the victim(s) and suspect(s) in terms of their blood sample with their consent. Hence, ~30–50% of the samples received in the forensic DNA laboratories constitute the whole blood samples. Additionally, for paternity dispute cases blood samples from the alleged father, mother, and child and in cases involving identification of mutilated remains reference blood sample from close blood relatives are generally sent for DNA examination. Whole blood is considered to be the most suitable target for isolation of nuclear DNA for forensic DNA analysis due to its high DNA content. Human blood is constituted of solid phase of formed elements, i.e., erythrocytes (RBC), leukocytes (WBC), and thrombocytes (platelets) and the plasma. Out of all the constituents only WBC are the nucleated cells containing the nuclear DNA. Thus, the amount of nuclear DNA in blood is determined by the amount of WBCs in a person's blood. WBC count varies greatly with the stages of immunosuppression, inflammation, or leukemia of the individual affecting the DNA content substantially. The extractable amount of DNA varies greatly with the protocol used which varies between 30 and 60 ng/μl. Most of the laboratories receive whole blood in the form of either liquid blood in EDTA vials or blood spotted on FTA™ cards. If blood has been collected in EDTA vials, it should be refrigerated (not frozen) and reach the laboratory at the earliest under refrigerated conditions.

H. R. Dash et al., *Principles and Practices of DNA Analysis: A Laboratory Manual for Forensic DNA Typing*, Springer Protocols Handbooks, https://doi.org/10.1007/978-1-0716-0274-4_4, © Springer Science+Business Media, LLC, part of Springer Nature 2020

2 Principle

Different techniques are available for isolation of DNA from whole blood. However, most of the techniques follow the same principle of separation of WBCs from the whole blood sample and other cellular debris such as hemoglobin followed by chemical/physical lysis of cell and nuclear membrane of WBCs to release nuclear DNA and removal of proteins, carbohydrates, and RNAs from the solution. The preliminary aim of any DNA isolation technique is to yield a high quality and quantity of DNA that can be preserved for a longer period under recommended conditions. The principal steps of DNA isolation involve the lysis of cell and nucleus which is achieved by the combining effect of the enzymes and detergents. Sodium dodecyl sulfate (SDS) is the most common detergent used in this purpose, whereas proteinase-K is mostly used for enzymatic degradation of proteins. The commonly used chemical technique to remove proteins from the DNA involves its denaturation and extraction into the organic phase when phenol and chloroform are used. The DNA, insoluble in organic solution, resides in the aqueous phase and can be precipitated, washed, and air-dried. Finally it can be dissolved in aqueous buffer for long-term storage. The generalized principle of DNA isolation from a eukaryotic cell has been shown in Fig. 4.1.

3 Reagents Required and Their Role

3.1 Tris-EDTA

The combination of Tris and EDTA forms the major constituent of lysis buffers. Tris acts as the buffering force, whose major role is to maintain pH of the solution mostly at 8.0. The most common form of Tris, i.e., Tris(hydroxymethyl) aminomethane with p*K*a of 8.1, has its buffering capacity between pH 7 and 9. Tris buffers are generally temperature sensitive and hence should be stored at recommended temperature only. Tris also has an additional

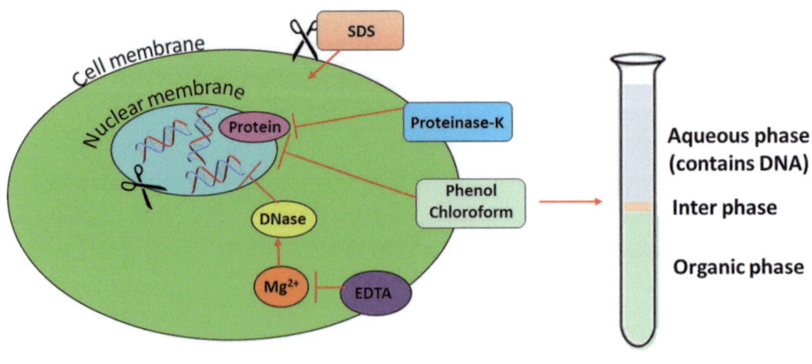

Fig. 4.1 Generalized principle of isolation of DNA from a eukaryotic cell

property of interacting with lipopolysaccharides (LPS) in the cell membrane, thus destabilizing the cell membrane to aid in cell lysis. Additionally, ethylenediaminetetraacetic acid (EDTA) binds to certain divalent cations mostly calcium (Ca^{2+}) and magnesium (Mg^{2+}). These divalent cations are reported to have their key role in the maintenance of membrane structure and function. Thus, eliminating these ions by EDTA leads to the lysis of cell membrane.

3.2 SDS

Sodium dodecyl sulfate (SDS) is a strong anionic detergent known to have protein denaturation properties. It helps to remove the negative ions from proteins to destabilize their structure and confirmation. In this process, SDS helps in rupturing the cell as well as nuclear membrane to release the genetic material into the solution. Additionally, it helps in releasing DNA from histone and other DNA-binding proteins. SDS is also known to have an additional property of removing lipid membranes, thus aiding in the lysis of cell and nuclear membrane.

3.3 Proteinase-K

Peptide bond present near the carboxyl group of amino acids with blocked alpha amino groups is the cleavage site for the serine protease, i.e., proteinase-K. It helps in degrading many protein impurities to yield a good quality of DNA. It also plays an important role in the inactivation of nucleases, hence preventing the isolated DNA from damage. The activity of proteinase-K is greatly increased by its simultaneous action with SDS and high temperature. Proteinase-K works optimally at 50–65 °C as higher temperature unfolds certain proteins to ease the function of the enzyme.

3.4 Phenol

In most of the cases, phenol is used in Tris-saturated form. Phenol, being a weak acid, is equilibrated with buffers to make it slightly alkaline which is the most suitable pH for DNA isolation. These buffer-saturated phenols contain around 72% of phenol with a slightly higher density than that of water. Phenol has the property to denature proteins during their presence in the aqueous solution. During this process, the nonaqueous component, i.e., protein, gets separated from the aqueous component, i.e., DNA.

3.5 Chloroform-Isoamyl Alcohol

This is mostly used at a ratio of 24:1. These are detergents that bind to the proteins and lipids to dissolve them in the organic solution. After dissolving, the solutions help in clumping the protein-lipid complexes to form a precipitate. Thus, the liquid-liquid extraction technique involving these chemicals distinguishes between the physical positions of the cell components, i.e., upper aqueous phase containing nucleic acids, lower organic phase containing protein components, and middle phase constituting the lipids.

3.6 Sodium Acetate

Salts like sodium or potassium acetate play an important role in neutralizing the charges on sugar phosphate backbone of DNA, e.g., sodium acetate breaks into Na^+ and CH_3COO^- in solution. The positive charge of Na^+ neutralizes the negative charge of PO_3^- groups of DNA to make it less hydrophilic and thus less soluble in water, which leads to dropout of DNA in the solution.

3.7 Isopropyl Alcohol

Alcohol precipitation is the most commonly used technique for desalting DNA followed by its precipitation. DNA is highly insoluble in isopropanol. Isopropanol dissolves in water to form a solution that causes the DNA in the solution to aggregate and precipitate. Additionally, isopropanol is used as a better alternative for ethanol due to its greater potential for DNA precipitation in lower concentrations. Besides, it takes lesser time to evaporate.

3.8 TE Buffer

TE buffer is generally prepared by mixing 50 mM Tris and 50 mM EDTA at pH 8.0. Tris, the major constituent of TE buffer, acts as a common pH buffer. Additionally, EDTA chelates cations like Mg^{2+}. Hence, TE buffer helps to solubilize DNA by protecting it from degradation.

4 Procedure

1. Take 500 μl of fresh blood or blood preserved in EDTA solution in a sterilized 2 ml microcentrifuge tube (MCT). Mark the tube carefully, and cover the marking with cello tape.
2. Add 500 μl of lysis buffer I to the tube containing 500 μl blood, cover the tube by Parafilm®, and mix thoroughly by vortex. Freeze the tubes at −70 °C for overnight.
3. Take out the sample carefully by keeping it on ice bucket and incubate the samples immediately in water bath earlier set at 65 °C for 10 min (instant heat shock of temperature difference ~100 °C lyse the cells).
4. Carefully take out the samples from water bath, and centrifuge at 10,000 rpm for 10 min at 4 °C; discard the supernatant carefully (at last a jelly-type substance remains).
5. Add 500 μl of lysis buffer II to the pellet, and mix carefully and properly by vortex. You may use the pipette tip to disrupt the pellet. Mix till complete mixing of the pellet; otherwise the yield of DNA may be reduced.
6. Add proteinase-K to make its final concentration as 100 μg/ml in total volume.
7. Add sodium dodecyl sulfate (SDS) to make its final concentration as 2% in total volume.

8. Cover the cap of the tube in Parafilm®, mix thoroughly, and incubate in water bath at 56 °C for 2 h.
9. After 2 h, take the tubes out of the water bath, add 500 μl (1:1 volume) of Tris-saturated phenol (TSP, pH 8.0), and mix thoroughly for 10 min by gentle shaking.
10. After proper mixing, centrifuge the tube at 10,000 rpm for 10 min at 4 °C.
11. Collect the upper aqueous phase by careful pipetting in a separate MCT (you may use cut tips).
12. Add half of the volume of TSP (250 μl in this case) and chloroform:isoamyl alcohol (24:1) (250 μl in this case), and mix thoroughly for 10 min by gentle shaking.
13. After proper mixing, centrifuge the tube at 10,000 rpm for 10 min at 4 °C. Transfer the upper aqueous layer to a fresh MCT.
14. Add 500 μl (1:1 volume) of chloroform:isoamyl alcohol (24:1), and mix thoroughly for 10 min by gentle shaking.
15. After proper mixing, centrifuge the tube at 10,000 rpm for 10 min at 4 °C. Transfer the upper aqueous supernatant to a fresh MCT.
16. Add 1/30th volume of 3 M sodium acetate (pH 5.5) and equal volume of chilled isopropyl alcohol.
17. DNA will be precipitated slowly; remove the reagents except the DNA pellet carefully from the MCT.
18. Wash the DNA pellet three times with 100 μl of 70% ethanol.
19. Vacuum dry the pellet and dissolve the pellet in 50–100 μl of TE buffer depending on the size of the pellet.
20. Incubate the tube at 56 °C for 10 min to remove RNA contamination.
21. Store the DNA at −20 °C for further use.

5 Observation

The quantity of the isolated DNA can be determined at preliminary level by UV–visible spectrophotometer. For a 1 cm path length, the optical density at 260 nm (OD_{260}) equals 1.0 for the following samples:

- 50 μg/ml of double-stranded DNA
- 33 μg/ml of single-stranded DNA
- 20–30 μg/ml of oligonucleotide
- 40 μg/ml of RNA

6 Result Table

Sample	DNA content	$OD_{260/280}$	Inference
Control			
Sample 1			
Sample 2			

OD optical density

7 Precautions

- Wear gloves and goggles during manual DNA isolation procedure.
- The plasticware to be used during the entire procedure should be pre-sterilized and DNase free.
- Use low-retention plasticware for maximum recovery of the samples.
- Handle the chemicals carefully, mostly the phenol and chloroform. Phenol, being a strong acid, may cause severe burns whereas chloroform is a known carcinogen.
- Use wide-bore tips by cutting 2–3 mm from the rear end to prevent mechanical disruption of DNA during pipetting.
- Use filter tips to avoid cross-sample contamination.
- Do not use the TSP solution if the color of the solution has changed to pink (occurs due to oxidation of phenol). This may promote nicks in the extracted DNA.
- Pipetting should be done carefully without mixing the two phases, i.e., upper aqueous phase and lower organic phase, at any point of time.
- From forensic DNA typing point of view match the name of the donor, date, crime no., and other details on the tube containing blood prior to its consumption for DNA isolation.

8 Troubleshooting

Problem	Tentative cause	Possible solutions
Protein contamination	More amount of initial cells (WBC) in the blood	Dilute the blood with normal saline solution prior to the initial step
	The extraction step has not been followed properly	Repeat the phenol: chloroform:isoamyl alcohol extraction step
	Blood in coagulated condition	Check the blood for its quality prior to the extraction steps
RNA contamination	High-temperature incubation not given to the extracted DNA	Incubate the extracted DNA at 56 °C for 10 min
	RNase not added	Add RNase at a final concentration of 400 μg/ml to the isolated DNA sample
No visible DNA pellet obtained	DNA quantity very less	Incubate the tubes at −20 °C for overnight after the last step of adding 3 M sodium acetate and chilled isopropanol
		Repeat the isolation procedure with increased amount of initial blood quantity
		Increase the concentration of proteinase-K and extend the time of incubation at 56 °C
Insoluble pellet after precipitation	Improper drying of the DNA precipitate	Avoid extended drying under vacuum conditions; DNA being acid in nature use a slightly alkaline solution such as TE pH 8.0 to dissolve the precipitate
Degraded DNA	Condition of blood not appropriate	Isolate DNA from coagulated blood by treating them as forensic samples
	Blood found coagulated	
	Blood stored for longer time	

Chapter 5

Manual Isolation of DNA from Body Fluid Stains

Objective: To isolate DNA from body fluid stains on clothing.

1 Introduction

Human body fluids are considered to be the major pieces of evidence from forensic analysis point of view as their presence at a particular location provides useful information to the investigators. With the advent of DNA technology, the usefulness of body fluid stains at a crime scene has increased further as DNA typing of these samples can not only ascertain a suspect, but also exonerate an innocence. Additionally, the presence of body fluids provides an indication regarding the sequential events unfolded during the crime, e.g., presence of bloodstain signifying struggle or murder, whereas the detection of seminal and/or vaginal fluid indicating sexual assault.

Biological examination of body fluids involves identification of body fluids by chemical, enzymatic, and microscopic examination followed by retrieving the biological materials by either swabbing, cutting, or scraping. Later, DNA analysis is performed with these samples to confirm the individual origin of the samples. From forensic point of view body fluids such as blood can be expected to be present at crime scene, weapon, and clothing of both deceased and suspect(s). In this regard, obtaining a consistent DNA profile from the objects flooded with bloodstains received from crime scene, deceased, weapon, and suspect(s) can solve a case of murder. Additionally, for the cases involving sexual assault, seminal stains can be expected at crime scene, victim's belongings, and swab of victim's private parts. In such cases, male DNA profile obtained from the victim's belongings can be matched with that of the suspect(s) to prove the assault. The prerequisite of all the

H. R. Dash et al., *Principles and Practices of DNA Analysis: A Laboratory Manual for Forensic DNA Typing*, Springer Protocols Handbooks, https://doi.org/10.1007/978-1-0716-0274-4_5, © Springer Science+Business Media, LLC, part of Springer Nature 2020

downstream processing techniques is the isolation of DNA from these body fluid stains which has been discussed in much details in this chapter.

Prior to the isolation of DNA from body fluid stains, the exact nature of origin of the body fluid needs to be ascertained. Selection of the preliminary examination to be conducted depends on the nature of the sample and the examiner's convenience. Some of the presumptive tests of detection of bloodstains include the oxidation-reduction reactions and colorimetric assays such as phenolphthalein assay, leucomalachite green (LMG) assay, and benzidine assay, whereas the confirmatory assays involve microcrystal assays and other chromatographic and electrophoretic techniques. Similarly, the preliminary tests to be conducted to ascertain seminal fluid involves acid phosphatase test, testing of antigens specific to prostate, seminal vesicle, examination under UV light, or confirmation of the presence of spermatozoa by microscopic observation.

2 Principle

Dried or wet bloodstains constitute various components of blood itself along with certain additional chemicals such as humic acid (if stain is collected from soil or floor) or chemicals constituting the dye of the clothing on which the stains are present. In a similar fashion seminal fluids are found along with the mixture of vaginal fluids and chemicals such as humic acids and dyes. In both the cases, DNA isolation technique follows the same principle as that of the DNA isolation from liquid blood by phenol-chloroform extraction method. The additional step involves the extraction of biological materials to the liquid medium from the solid adherent surface. In this process, the clothing harboring body fluid stains can be incubated along with buffer for some time followed by separation of clothing by squeezing to release the body fluid materials present in the form of stains into the solution.

Most of the protocols vary in terms of their buffer pH, chemical compositions, time, and temperature; however, the principle of DNA isolation by organic extraction method remains the same for all the samples, i.e., application of SDS to lyse the cell wall followed by the use of proteinase-K to digest proteins and liquid-liquid extraction method using phenol-chloroform-isoamyl alcohol. As sperm cells are more robust, an additional component dithiothreitol (DTT) is applied during the step of cell lysis to release the DNA into the solution.

3 Reagents Required and Their Role

3.1 Tris-EDTA

The combination of Tris and EDTA forms the major constituent of lysis buffers. Tris acts as the buffering force, whose major role is to maintain pH of the solution mostly at 8.0. The most common form of Tris, i.e., Tris(hydroxymethyl) aminomethane with p*Ka* of 8.1, has its buffering capacity between pH 7 and 9. Tris buffers are generally temperature sensitive and hence should be stored at recommended temperature only. Tris also has an additional property of interacting with lipopolysaccharides (LPS) in the cell membrane, thus destabilizing the cell membrane to aid in cell lysis. Additionally, ethylenediaminetetraacetic acid (EDTA) binds to certain divalent cations mostly calcium (Ca^{2+}) and magnesium (Mg^{2+}). These divalent cations are reported to have their key role in the maintenance of membrane structure and function. Thus, eliminating these ions by EDTA leads to the lysis of cell membrane.

3.2 SDS

Sodium dodecyl sulfate (SDS) is a strong anionic detergent known to have protein denaturation properties. It helps to remove the negative ions from proteins to destabilize their structure and confirmation. In this process, SDS helps in rupturing the cell as well as nuclear membrane to release the genetic material into the solution. Additionally, it helps in releasing DNA from histone and other DNA-binding proteins. SDS is also known to have an additional property of removing lipid membranes and thus aiding in the lysis of cell and nuclear membrane.

3.3 Proteinase-K

Peptide bond present near the carboxyl group of amino acids with blocked alpha amino groups is the cleavage site for the serine protease, i.e., proteinase-K. It helps in degrading many protein impurities to yield a good quality of DNA. It also plays an important role in the inactivation of nucleases, hence preventing the isolated DNA from damage. The activity of proteinase-K is greatly increased by its simultaneous action with SDS and high temperature. Proteinase-K works optimally at 50–65 °C as higher temperature unfolds certain proteins to ease the function of the enzyme.

3.4 DTT

Dithiothreitol (DTT) prevents the formation of intramolecular and intermolecular disulfide bonds among cysteine residues. With protein disulfide bridges being the major constituent of sperm nuclear membrane, DTT is essential during lysis of sperm cells to release sperm DNA. Hence, during the lysis of sperm cells, in addition to proteinase-K and SDS, DTT is also used in the mixture for their effective lysis.

3.5 Phenol

In most of the cases, phenol is used in Tris-saturated form. Phenol, being a weak acid, is equilibrated with buffers to make it slightly alkaline which is the most suitable pH for DNA isolation. These buffer-saturated phenols contain around 72% of phenol with a slightly higher density than that of water. Phenol has the property to denature proteins during their presence in the aqueous solution. During this process, the nonaqueous component, i.e., protein, gets separated from the aqueous component, i.e., DNA.

3.6 Chloroform-Isoamyl Alcohol

This is mostly used at a ratio of 24:1. These are detergents that bind to the proteins and lipids to dissolve them in the organic solution. After dissolving, the solutions help in clumping the protein-lipid complexes to form a precipitate. Thus, the liquid-liquid extraction technique involving these chemicals distinguishes between the physical positions of the cell components, i.e., upper aqueous phase containing nucleic acids, lower organic phase containing protein components, and middle phase constituting the lipids.

3.7 TE Buffer

TE buffer is generally prepared by mixing 50 mM Tris and 50 mM EDTA at pH 8.0. Tris, the major constituent of TE buffer, acts as a common pH buffer. Additionally, EDTA chelates cations like Mg^{2+}. Hence, TE buffer helps to solubilize DNA by protecting it from degradation.

4 Procedure

1. Cut the clothing containing suspected stains into small pieces of 0.5 cm × 0.5 cm, place them in a properly labeled sterilized 2 ml microcentrifuge tube (MCT), and cover the labeling with cello tape.
2. Add 500 μl of forensic buffer to the tube containing stain pieces, and mix thoroughly by vortex.
3. Incubate the samples at 37 °C for 3–4 h.
4. Squeeze the clothing into the same tube with the help of a disposable syringe; now the tube contains the extracts of the stains in 500 μl of forensic buffer.
5. Add proteinase-K to make its final concentration as 100 μg/ml in total volume [in this case, add 2.5 μl from the 20 mg/ml stock solution].
6. Add sodium dodecyl sulfate (SDS) to make its final concentration as 2% in total volume [in this case, add 500 μl from the 20% stock solution].
7. If the extract contains suspected seminal fluid, optionally add DTT as 150 mM in total volume.
8. Cover the cap of the tube in Parafilm®, mix thoroughly, and incubate in water bath at 56 °C for 2–4 h.

9. After 2–4 h, take the samples out of the water bath, add 500 μl (1:1 volume) of Tris-saturated phenol (TSP, pH 8.0), and mix thoroughly for 10 min by gentle shaking.
10. After proper mixing, centrifuge the tube at 10,000 rpm for 10 min at 4 °C.
11. Collect the upper aqueous phase by careful pipetting in a separate MCT (you may use cut tips).
12. Add half of the volume of TSP (250 μl in this case) and chloroform:isoamyl alcohol (24:1) (250 μl in this case), and mix thoroughly for 10 min by gentle shaking.
13. After proper mixing, centrifuge the tube at 10,000 rpm for 10 min at 4 °C. Transfer the upper aqueous layer to a fresh MCT.
14. Add 500 μl (1:1 volume) of chloroform:isoamyl alcohol (24:1), and mix thoroughly for 10 min by gentle shaking.
15. After proper mixing, centrifuge the tube at 10,000 rpm for 10 min at 4 °C.
16. Keep an Amicon® Ultra-0.5 ml 30 kDa centrifugal filters into a 2.0 ml MCT.
17. Transfer the upper aqueous supernatant to the filter device and cap it.
18. Spin the device at 5000 × *g* for 10 min at 20 °C. Discard the flow through.
19. Add 500 μl of Milli-Q to the device for washing and repeat spinning of the device at 5000 × *g* for 10 min at 20 °C.
20. Remove the assembled device and to recover the concentrated solute place the Amicon® Ultra-0.5 ml 30 kDa centrifugal filter in a fresh MCT upside down.
21. Counterbalance with a similar device and spin at 1000 × *g* for 2 min at 20 °C.
22. Store the DNA at −20 °C for further use.

5 Observation

The quantity of the isolated DNA can be determined at preliminary level by UV–visible spectrophotometer. For a 1 cm path length, the optical density at 260 nm (OD_{260}) equals 1.0 for the following samples:

- 50 μg/ml of double-stranded DNA
- 33 μg/ml of single-stranded DNA
- 20–30 μg/ml of oligonucleotide
- 40 μg/ml of RNA

6 Result Table

Sample	DNA content	$OD_{260/280}$	Inference
Control			
Sample 1 (*bloodstain*)			
Sample 2 (*seminal stain*)			

OD optical density

7 Precautions

- Wear gloves and goggles during manual DNA isolation procedure.
- The plasticware to be used during the entire procedure should be sterilized and DNase free.
- Handle the chemicals carefully, mostly the phenol and chloroform. Phenol, being a strong acid, may cause severe burns whereas chloroform is a known carcinogen.
- Use wide-bore tips by cutting 2–3 mm from the rear end to prevent mechanical disruption of DNA during pipetting.
- Use filter tips to avoid cross-sample contamination.
- Do not use the TSP solution if the color of the solution has changed to pink (occurs due to oxidation of phenol). This may promote nicks in the extracted DNA.
- Pipetting should be done carefully without mixing the two phases, i.e., upper aqueous phase and lower organic phase, at any point of time.
- Perform the preliminary and confirmatory tests of the body fluids prior to the DNA isolation from them.
- Do not overincubate the stained clothing for stain extraction. It may extract dyes from the clothing to the extract.
- Note down the pattern of stains found on the forensic evidence before cutting them.

8 Troubleshooting

Problem	Tentative cause	Possible solutions
Protein contamination	The extraction step has not been followed properly	Repeat the phenol:chloroform:isoamyl alcohol extraction step
RNA contamination	High-temperature incubation not given to the extracted DNA	Incubate the extracted DNA at 56 °C for 10 min
	RNase not added	Add RNase at a final concentration of 400 μg/ml to the isolated DNA sample
Colored DNA extract	Extended incubation during extraction of stains	Avoid extended incubation during stain extraction
	Nature of the clothing	Use alternative DNA isolation methods such as magnetic bead-based extraction

Chapter 6

Manual Isolation of DNA from Soft-Tissue Samples

Objective: To isolate DNA from soft-tissue samples.

1 Introduction

Human tissues, either preserved or dried, are the most common biological evidence for DNA examination. The most common among the tissue samples received at laboratories include the fetus of early stage of development or the product of conception aborted illegally. Other forms of tissue samples received include tissue fragments attached to skeletal remains, formalin-preserved organs and tissues, and paraffin blocks containing tissues collected earlier for pathological diagnostics. During autopsy, viscera of the deceased are collected for toxicological analysis, which can also be used for DNA examination during nonavailability of other samples if stored and preserved properly.

Soft tissues are recommended to be preserved at −80 °C after their collection, failure of which may affect the quality as well as quantity of DNA necessary for DNA analysis. Additionally, putrefaction of human corpse also affects adversely the DNA content of the preserved tissues which is a major challenge in many varying climatic conditions. Other factors contributing to the DNA analysis of human corpse from soft tissues include mutilation of skeletal remains in runover situations, contamination of mutilated remains, and exposure of corpse to extreme environmental conditions such as mass disaster.

Formaldehyde is a common preservative for histopathological analysis of tissue samples. However, formalin can affect adversely the DNA content of a cell by either fragmentation, nucleotide base modification, or cross-linking between DNA and protein or DNA itself. Such fragmented highly damaged DNA poses a huge challenge for DNA typing experiments. Though many protocols have been developed for isolation of DNA from formalin-fixed tissues,

H. R. Dash et al., *Principles and Practices of DNA Analysis: A Laboratory Manual for Forensic DNA Typing*, Springer Protocols Handbooks, https://doi.org/10.1007/978-1-0716-0274-4_6, © Springer Science+Business Media, LLC, part of Springer Nature 2020

still it is not considered as a suitable preservative for tissue samples to be used for forensic DNA typing. In this regard, sterilized normal saline solution is recommended for tissue sample preservation. Optionally antibiotics (e.g., 0.1 mg/ml gentamicin solution) can also be added to the normal saline solution for proper preservation of soft-tissue samples. Tissue samples can also be stored without any preservative at −70 °C till further use.

2 Principle

Fresh soft tissues contain abundant cellular materials of parenchymal cells which are a huge source of DNA. Certain types of soft tissues such as cervical lymph nodes and muscle are the preferred samples for DNA extraction. These fresh tissues should be excised using sterile scalpel from fresh cadavers followed by their homogenization using a sterilized mill to prevent cross contamination. However, utmost precaution should be taken while dealing with putrefied soft-tissue samples. These samples may contain a huge amount of DNA, but care should be taken to eliminate the PCR inhibitors from these kind of samples.

The first step during DNA fingerprinting analysis involves the isolation of DNA from tissue sample. Though many commercial kits and technologies are available for isolation of DNA from tissue samples, traditional organic extraction method has a huge advantage over other techniques. In this technique, tissue samples are disaggregated by physical and chemical techniques followed by cell lysis using detergents and proteinase-K for protein digestion. Later, a liquid-liquid extraction method is employed by using phenol, chloroform, and isoamyl alcohol targeting differential solubility of proteins, lipids, and nucleic acids. Finally, DNA present in the aqueous phase is concentrated as well as purified by the use of salt and ethanol, followed by its resolubilization in Tris-EDTA buffer.

3 Reagents Required and Their Role

3.1 Tris-EDTA

The combination of Tris and EDTA forms the major constituent of lysis buffers. Tris acts as the buffering force, whose major role is to maintain the pH of the solution mostly at 8.0. The most common form of Tris, i.e., Tris(hydroxymethyl) aminomethane with p*Ka* of 8.1, has its buffering capacity between pH 7 and 9. Tris buffers are generally temperature sensitive and hence should be stored at recommended temperature only. Tris also has an additional property of interacting with lipopolysaccharides (LPS) in the cell membrane, thus destabilizing the cell membrane to aid in cell lysis. Additionally, ethylenediaminetetraacetic acid (EDTA) binds to certain divalent cations mostly calcium (Ca^{2+}) and magnesium (Mg^{2+}). These divalent cations are reported to have their key role in the maintenance of membrane structure and function. Thus, eliminating these ions by EDTA leads to the lysis of cell membrane.

3.2 SDS

Sodium dodecyl sulfate (SDS) is a strong anionic detergent known to have protein denaturation properties. It helps to remove the negative ions from proteins to destabilize their structure and confirmation. In this process, SDS helps in rupturing the cell as well as nuclear membrane to release the genetic material into the solution. Additionally, it helps in releasing DNA from histone and other DNA-binding proteins. SDS is also known to have an additional property of removing lipid membranes, thus aiding in the lysis of cell and nuclear membrane.

3.3 Proteinase-K

Peptide bond present near the carboxyl group of amino acids with blocked alpha amino groups is the cleavage site for the serine protease, i.e., proteinase-K. It helps in degrading many protein impurities to yield a good quality of DNA. It also plays an important role in the inactivation of nucleases, hence preventing the isolated DNA from damage. The activity of proteinase-K is greatly increased by its simultaneous action with SDS and high temperature. Proteinase-K works optimally at 50–65 °C as higher temperature unfolds certain proteins to ease the function of the enzyme.

3.4 Liquid Nitrogen

Liquid nitrogen is the prime requirement during mechanical homogenization of the tissue samples by cryogenic grinding. During this process, the tissue samples are snap frozen by dropping them on a beaker of liquid nitrogen. After grinding, the samples turn into small particle generating higher surface area for the action of chemical reagents for lysis of cell wall. However, the disadvantage of using liquid nitrogen-based grinding is the chance of sample loss during grinding of low-quantity samples.

3.5 Phenol

In most of the cases, phenol is used in Tris-saturated form. Phenol, being a weak acid, is equilibrated with buffers to make it slightly alkaline which is the most suitable pH for DNA isolation. These buffer-saturated phenols contain around 72% of phenol with a slightly higher density than that of water. Phenol has the property to denature proteins during their presence in the aqueous solution. During this process, the nonaqueous component, i.e., protein, gets separated from the aqueous component, i.e., DNA.

3.6 Chloroform-Isoamyl Alcohol

This is mostly used at a ratio of 24:1. These are detergents that bind to the proteins and lipids to dissolve them in the organic solution. After dissolving, the solutions help in clumping the protein-lipid complexes to form a precipitate. Thus, the liquid-liquid extraction technique involving these chemicals distinguishes between the physical positions of the cell components, i.e., upper aqueous phase containing nucleic acids, lower organic phase containing protein components, and middle phase constituting the lipids.

3.7 TE Buffer

TE buffer is generally prepared by mixing 50 mM Tris and 50 mM EDTA at pH 8.0. Tris, the major constituent of TE buffer, acts as a common pH buffer. Additionally, EDTA chelates cations like Mg^{2+}. Hence, TE buffer helps to solubilize DNA by protecting it from degradation.

4 Procedure

1. Cut solid tissue samples of size 3 mm^3 into small pieces of around 1–2 mm using sterile scalpel.
2. Either process these tissue samples immediately within 2 h or store them at −80 °C till further use.
3. Put the tissue samples in a clean mortar and dip it with liquid nitrogen. Using a clean pestle, grind the frozen tissue into small pieces.
4. Allow liquid nitrogen to evaporate leaving dry frozen grinded tissue in the mortar.
5. Add 500 μl of forensic buffer to the mortar, containing tissue powder, and wait till the tissue powder to dissolve completely in the buffer.
6. Resuspend the samples in buffer completely by gentle pipetting and transfer it to a 2.0 ml MCT.
7. Add proteinase-K to make its final concentration as 100 μg/ml in total volume.
8. Add sodium dodecyl sulfate (SDS) to make its final concentration as 2% in total volume.
9. Cover the cap of the tube in Parafilm®, mix thoroughly, and incubate in water bath at 56 °C for 2–4 h.
10. After 2–4 h, take the samples out of the water bath, add 500 μl (1:1 volume) of Tris-saturated phenol (TSP, pH 8.0), and mix thoroughly for 10 min by gentle shaking.
11. After proper mixing, centrifuge the tube at 10,000 rpm for 10 min at 4 °C.
12. Collect the upper aqueous phase by careful pipetting in a separate MCT (you may use cut tips).
13. Add half of the volume of TSP (250 μl in this case) and chloroform:isoamyl alcohol (24:1) (250 μl in this case), and mix thoroughly for 10 min by gentle shaking.
14. After proper mixing, centrifuge the tube at 10,000 rpm for 10 min at 4 °C. Transfer the upper aqueous layer to a fresh MCT.
15. Add 500 μl (1:1 volume) of chloroform:isoamyl alcohol (24:1), and mix thoroughly for 10 min by gentle shaking.
16. After proper mixing, centrifuge the tube at 10,000 rpm for 10 min at 4 °C.

17. Keep an Amicon® Ultra-0.5 ml 30 kDa centrifugal filters into a 2.0 ml MCT.
18. Transfer the upper aqueous supernatant to the filter device and cap it.
19. Spin the device at 5000 × *g* for 10 min at 20 °C. Discard the flow through.
20. Add 500 μl of Milli-Q to the device for washing and repeat spinning of the device at 5000 × *g* for 10 min at 20 °C.
21. Remove the assembled device and to recover the concentrated solute place the Amicon® Ultra-0.5 ml 30 kDa centrifugal filter in a fresh MCT upside down.
22. Counterbalance with a similar device and spin at 1000 × *g* for 2 min at 20 °C.
23. Store the DNA at −20 °C for further use.

5 Observation

The quantity of the isolated DNA can be determined at preliminary level by UV–visible spectrophotometer. For a 1 cm path length, the optical density at 260 nm (OD_{260}) equals 1.0 for the following samples:

- 50 μg/ml of double-stranded DNA
- 33 μg/ml of single-stranded DNA
- 20–30 μg/ml of oligonucleotide
- 40 μg/ml of RNA

6 Result Table

Sample	DNA content	$OD_{260/280}$	Inference
Control			
Sample 1 (*fresh tissue*)			
Sample 2 (*decomposed tissue*)			

OD optical density

7 Precautions

- Wear gloves and goggles during manual DNA isolation procedure.
- The plasticware to be used during the entire procedure should be sterilized and DNase free.

- Handle the chemicals carefully, mostly the phenol and chloroform. Phenol, being a strong acid, may cause severe burns whereas chloroform is a known carcinogen.
- Use wide-bore tips by cutting 2–3 mm from the rear end to prevent mechanical disruption of DNA during pipetting.
- Use filter tips to avoid cross-sample contamination.
- Do not use the TSP solution if the color of the solution has changed to pink (occurs due to oxidation of phenol). This may promote nicks in the extracted DNA.
- Pipetting should be done carefully without mixing the two phases, i.e., upper aqueous phase and lower organic phase, at any point of time.
- During grinding, cover the mouth of the mortar using a paper cloth to avoid spillage of tissue fragments.
- While grinding, work inside a hood to protect yourself from aerosolized powder.
- Handle liquid nitrogen carefully.

8 Troubleshooting

Problem	Tentative cause	Possible solutions
Protein contamination	The extraction step has not been followed properly	Repeat the phenol: chloroform:isoamyl alcohol extraction step
RNA contamination	High-temperature incubation not given to the extracted DNA	Incubate the extracted DNA at 56 °C for 10 min
	RNase not added	Add RNase at a final concentration of 400 µg/ ml to the isolated DNA sample
PCR product not amplified	Presence of PCR inhibitors	Clean the samples properly; if possible wash the tissue samples with sterilized normal saline solution prior to their processing
Partial DNA profile obtained	Tissue stored in formalin	Do not process tissue samples stored in formalin, and use alternative protocols

Chapter 7

Manual Isolation of DNA from Rooted Hair Samples

Objective: To isolate DNA from rooted hair samples.

1 Introduction

Human hair originating from either scalp or pubic region is a great source of biological evidence for forensic implications. Earlier days, hair examination was compromising of their morphological and microscopic analysis for comparison. However, with the advent of PCR technology, the DNA content of hairs is regarded as the most useful source of biological evidence from crime scene. The small quantity of DNA present in the hair has been highly acknowledged for their application in individualization.

A typical human hair has two parts, i.e., root and shaft. Root anchors the hair on the skin whereas the shaft grows outside the skin surface to be seen outside. Medulla is the center position of hair which is surrounded by cortex and the whole hair is protected by layers of dead keratinized cells called as cuticle. Hairs are originated from the follicles which are located at dermis and epidermis. Many nerves and capillaries supplying nutrients open up at the base of the hair root known as papilla and are surrounded by hair bulb.

As any cell contains many copies of mitochondrial DNA, they can be successfully extracted from both hair roots and shafts. Subsequently, the sequence polymorphism of mitochondrial DNA can be analyzed to discriminate among individuals. However, as mitochondrial DNA is inherited matrilineally, they are not a suitable candidate for analysis in paternity testing as well as individualization with superior degree of resolution. Hence, analysis of nuclear DNA from the hairs is of great interest from forensic DNA analysis point of view. Freshly plucked hairs with roots are of great use in nuclear DNA analysis as cells of root region contain higher amount of such DNA. Hair shafts or naturally shed hairs contain very less DNA in

H. R. Dash et al., *Principles and Practices of DNA Analysis: A Laboratory Manual for Forensic DNA Typing*, Springer Protocols Handbooks, https://doi.org/10.1007/978-1-0716-0274-4_7, © Springer Science+Business Media, LLC, part of Springer Nature 2020

comparison to the DNA content of root in freshly plucked hairs. The average DNA content of a single hair and its surrounding cells is 0.5 μg. Most of the hairs found in crime scene are generally shed hairs, which contain very less amount of DNA. Thus, it is always recommended to use multiple number of hairs simultaneously for DNA isolation. Additionally, many studies have recovered nuclear DNA from hair shaft; however, the quantity is considerably less in comparison to the hair roots.

In spite of the presence of high quantity of nuclear DNA in anagenic (growth phase) hairs and presence of less quantity of DNA in telogenic (resting phase) hairs, hairs are always useful from forensic analysis point of view. In certain instances, hairs devoid of roots, i.e., shaft part found in crime scene, also yield sufficient amount of nuclear DNA due to the presence of blood on these hairs. Additionally, a common form of skin alteration, i.e., dandruff, can also be used as a source of DNA for DNA typing analysis as the quantity of DNA varies greatly from 0.8 to 16.6 ng per dandruff particle.

2 Principle

Hair follicles of plucked hairs contain certain amount of epithelial cells and hence contain a huge amount of DNA. Hairs can be transferred by direct contact, and struggle followed by their deposit on the clothing, on an object, or at the crime scene. Preliminary microscopic examination of the presence or absence of hair follicles should be performed prior to their consumption for DNA isolation. Hairs found on the objects such as hat, blanket, and pillow can also be targeted as forensic evidence. These samples should always be analyzed along with the reference samples for matching of DNA profile.

Though hair shaft contains few amount of DNA, isolation of DNA targets mostly the hair roots which contain a higher amount of DNA. In the first step of DNA fingerprinting analysis it is necessary to isolate sufficient quantity as well as quality of DNA from the hair samples. In traditional method of DNA isolation from hair samples, preliminarily cells are digested using detergents and proteinase-K for protein digestion-based cell disruption. Subsequently, the liquid-liquid extraction method is followed employing phenol, chloroform, and isoamyl alcohol targeting differential solubility of proteins, lipids, and nucleic acids. Lastly, DNA is concentrated and purified by using salt and ethanol. Optionally, resolubilization of the extracted DNA is performed in Tris-EDTA buffer.

3 Reagents Required and Their Role

3.1 Tris-EDTA

The combination of Tris and EDTA forms the major constituent of lysis buffers. Tris acts as the buffering force, whose major role is to maintain pH of the solution mostly at 8.0. The most common form of Tris, i.e., Tris(hydroxymethyl) aminomethane with p*K*a of 8.1, has its buffering capacity between pH 7 and 9. Tris buffers are generally temperature sensitive and hence should be stored at recommended temperature only. Tris also has an additional property of interacting with lipopolysaccharides (LPS) in the cell membrane, thus destabilizing the cell membrane to aid in cell lysis. Additionally, ethylenediaminetetraacetic acid (EDTA) binds to certain divalent cations mostly calcium (Ca^{2+}) and magnesium (Mg^{2+}). These divalent cations are reported to have their key role in the maintenance of membrane structure and function. Thus, eliminating these ions by EDTA leads to the lysis of cell membrane.

3.2 SDS

Sodium dodecyl sulfate (SDS) is a strong anionic detergent known to have protein denaturation properties. It helps to remove the negative ions from proteins to destabilize their structure and confirmation. In this process, SDS helps in rupturing the cell as well as nuclear membrane to release the genetic material into the solution. Additionally, it helps in releasing DNA from histone and other DNA-binding proteins. SDS is also known to have an additional property of removing lipid membranes, thus aiding in the lysis of cell and nuclear membrane.

3.3 Proteinase-K

Peptide bond present near the carboxyl group of amino acids with blocked alpha amino groups is the cleavage site for the serine protease, i.e., proteinase-K. It helps in degrading many protein impurities to yield a good quality of DNA. It also plays an important role in the inactivation of nucleases, hence preventing the isolated DNA from damage. The activity of proteinase-K is greatly increased by its simultaneous action with SDS and high temperature. Proteinase-K works optimally at 50–65 °C as higher temperature unfolds certain proteins to ease the function of the enzyme.

3.4 Phenol

In most of the cases, phenol is used in Tris-saturated form. Phenol, being a weak acid, is equilibrated with buffers to make it slightly alkaline which is the most suitable pH for DNA isolation. These buffer-saturated phenols contain around 72% of phenol with a slightly higher density than that of water. Phenol has the property to denature proteins during their presence in the aqueous solution. During this process, the nonaqueous component, i.e., protein, gets separated from the aqueous component, i.e., DNA.

3.5 Chloroform-Isoamyl Alcohol

This is mostly used at a ratio of 24:1. These are detergents that bind to the proteins and lipids to dissolve them in the organic solution. After dissolving, the solutions help in clumping the protein-lipid complexes to form a precipitate. Thus, the liquid-liquid extraction technique involving these chemicals distinguishes between the physical positions of the cell components, i.e., upper aqueous phase containing nucleic acids, lower organic phase containing protein components, and middle phase constituting the lipids.

3.6 TE Buffer

TE buffer is generally prepared by mixing 50 mM Tris and 50 mM EDTA at pH 8.0. Tris, the major constituent of TE buffer, acts as a common pH buffer. Additionally, EDTA chelates cations like Mg^{2+}. Hence, TE buffer helps to solubilize DNA by protecting it from degradation.

4 Procedure

1. Examine the hairs carefully (may be under microscope) for the presence of hair root.
2. Wash the hairs with sterilized deionized water to remove the debris associated with the hairs.
3. Take 2–3 rooted hairs, cut 1 cm portion of the hair from the root end, and place it in a labeled microcentrifuge tube.
4. Add 500 μl of hair digestion buffer (for composition see the Appendix section), and make sure that the hairs are dipped inside the buffer.
5. Add proteinase-K to make its final concentration as 200 μg/ml in total volume.
6. Cover the cap of the tube in Parafilm®, mix thoroughly, and incubate in water bath at 56 °C for 2–4 h.
7. After 2–4 h, take the samples out of the water bath, add 500 μl (1:1 volume) of Tris-saturated phenol (TSP, pH 8.0), and mix thoroughly for 10 min by gentle shaking.
8. After proper mixing, centrifuge the tube at 10,000 rpm for 10 min at 4 °C.
9. Collect the upper aqueous phase by careful pipetting in a separate MCT (you may use cut tips).
10. Add half of the volume of TSP (250 μl in this case) and chloroform:isoamyl alcohol (24:1) (250 μl in this case), and mix thoroughly for 10 min by gentle shaking.
11. After proper mixing, centrifuge the tube at 10,000 rpm for 10 min at 4 °C. Transfer the upper aqueous layer to a fresh MCT.

12. Add 500 μl (1:1 volume) of chloroform:isoamyl alcohol (24:1), and mix thoroughly for 10 min by gentle shaking.
13. After proper mixing, centrifuge the tube at 10,000 rpm for 10 min at 4 °C.
14. Keep an Amicon® Ultra-0.5 ml 30 K centrifugal filters into a 2.0 ml MCT.
15. Transfer the upper aqueous supernatant to the filter device and cap it.
16. Spin the device at 5000 × *g* for 10 min at 20 °C. Discard the flow through.
17. Add 500 μl of Milli-Q to the device for washing and repeat spinning of the device at 5000 × *g* for 10 min at 20 °C.
18. Remove the assembled device and to recover the concentrated solute place the Amicon® Ultra-0.5 ml 30 K centrifugal filter in a fresh MCT upside down.
19. Counterbalance with a similar device and spin at 1000 × *g* for 2 min at 20 °C.
20. Store the DNA at −20 °C for further use.

5 Observation

The quantity of the isolated DNA can be determined at preliminary level by UV–visible spectrophotometer. For a 1 cm path length, the optical density at 260 nm (OD_{260}) equals 1.0 for the following samples:

- 50 μg/ml of double-stranded DNA
- 33 μg/ml of single-stranded DNA
- 20-30 μg/ml of oligonucleotide
- 40 μg/ml of RNA

6 Result Table

Sample	DNA content	$OD_{260/280}$	Inference
Control			
Sample 1 (*hair root*)			
Sample 2 (*hair shaft*)			

OD optical density

7 Precautions

- Wear gloves and goggles during manual DNA isolation procedure.
- The plasticware to be used during the entire procedure should be sterilized and DNase free.
- Handle the chemicals carefully, mostly the phenol and chloroform. Phenol, being a strong acid, may cause severe burns whereas chloroform is a known carcinogen.
- Use wide-bore tips by cutting 2–3 mm from the rear end to prevent mechanical disruption of DNA during pipetting.
- Use filter tips to avoid cross-sample contamination.
- Do not use the TSP solution if the color of the solution has changed to pink (occurs due to oxidation of phenol). This may promote nicks in the extracted DNA.
- Pipetting should be done carefully without mixing the two phases, i.e., upper aqueous phase and lower organic phase, at any point of time.
- Check the presence or absence of root hair prior to the isolation procedure.
- Check for the presence/absence of any body fluids associated with the hairs prior to DNA isolation.
- Do not extract DNA from hair samples and reference blood samples simultaneously.

8 Troubleshooting

Problem	Tentative cause	Possible solutions
Protein contamination	The extraction step has not been followed properly	Repeat the phenol: chloroform:isoamyl alcohol extraction step
RNA contamination	High-temperature incubation not given to the extracted DNA	Incubate the extracted DNA at 56 °C for 10 min
	RNase not added	Add RNase at a final concentration of 400 μg/ml to the isolated DNA sample
PCR product not amplified	Presence of PCR inhibitors	Clean the samples properly; if possible wash the hairs repeatedly with sterilized deionized water prior to their processing

(continued)

Problem	Tentative cause	Possible solutions
Mixed DNA profile obtained	Presence of body fluids along with hairs	Wash the samples properly before the isolation procedure
	Hairs originating from more than one individual	Process individual hair samples separately

Chapter 8

Manual Isolation of DNA from Human Teeth

Objective: To isolate DNA from tooth samples.

1 Introduction

Human teeth are chosen to be the most suitable part of highly degraded human remains for forensic identification purposes. Human teeth have a unique composition and their location within the jawbone protects them from environmental degradations over the years. DNA obtained from the tooth samples is of superior quantity and quality in comparison to bone samples. Additionally, DNA is preserved in teeth for a very long time and becomes a useful tool for ancient DNA examination of samples 100–10,000 years old.

Human embryonic development gives rise to two types of teeth, i.e., 20 numbers of deciduous or primary teeth, which are later replaced by 32 numbers of secondary or permanent dentition. The major structure of a tooth is composed of the mineralized matrix called as dentin which does not contain any cells. The crown of the dentin is covered by a layer of tooth enamel constituted of crystalline form of calcium phosphate. The interior chamber of tooth, i.e., central pulp cavity, is connected with the nerves and blood vessels through the root canal. It is the major source of DNA in tooth samples. Human teeth have been divided into four types, i.e., incisors, canines, premolars, and molars. Both incisors and canines constitute single roots, whereas premolars have one or two roots. However, the molars have three or more roots, thus projecting them to be a huge potential source of DNA for further analysis.

Characterization of teeth and their alignment inside mouth provide a useful information regarding the identification of an unknown person from radiographic and biochemical point of

H. R. Dash et al., *Principles and Practices of DNA Analysis: A Laboratory Manual for Forensic DNA Typing*, Springer Protocols Handbooks, https://doi.org/10.1007/978-1-0716-0274-4_8, © Springer Science+Business Media, LLC, part of Springer Nature 2020

view. From forensic dentistry point of view, a victim or accused can be connected by the analysis of dental remains such as X-rays and dental casts as well as the analysis of bite marks. However, DNA fingerprinting is the most useful tool for forensic analysis of tooth in cases of nonavailability of dental records or suitable controls for comparison. A good quality and quantity of DNA are expected from the dental pulp tissue. Hence, it is always recommended to collect rooted, intact teeth mostly with multiple teeth roots, which can be preserved and transported at room temperature.

2 Principle

DNA is primarily present at the inner surface, i.e., dental pulp of the teeth. Hence, the preliminary step during isolation of DNA from tooth samples is to break open the teeth to expose their inner surface. In this context, various steps have been exploited starting from sectioning of teeth at cementoenamel junction horizontally or vertically up to the root tip. Few other techniques involve the crushing, cryogenic grinding, or retrieval of dental pulp by various techniques. The technique involved should be examined carefully for minimal/no introduction of contaminants as well as inhibitors. Other environmental factors such as time, temperature, humidity, light, and other chemical substances should also be assessed prior to the isolation of DNA from the tooth sample.

DNA isolation of tooth sample involves three different stages, i.e., lysis of the cells followed by denaturation and inactivation of proteins and finally extraction of DNA. Though many commercial kits and technologies are available for isolation of DNA from tooth samples, traditional organic extraction method has a huge advantage over other techniques. The tooth samples with exposed root canals are subjected to lysis using detergents and proteinase-K by protein digestion. Later, a liquid-liquid extraction method is employed by using phenol, chloroform, and isoamyl alcohol targeting differential solubility of proteins, lipids, and nucleic acids. Finally, DNA present in the aqueous phase is concentrated as well as purified by the use of salt and ethanol, followed by its resolubilization in Tris-EDTA buffer.

3 Reagents Required and Their Role

3.1 Sodium Hypochlorite

Sodium hypochlorite (NaOCl) is the most widely used disinfectant and is known for its excellent cleaning action. However, the effectiveness of the solution is dependent on the concentration and pH of the solution. pH of the solution effects the formation of the hypochlorite ion (^{-}OCl) and proton (H^{+}) by dissociation of the weak acid hypochlorous acid (HOCl). In this case, HOCl plays an important role in germicidal action whereas the $^{-}$OCl determines the cleansing efficacy of the solution.

3.2 Tris-EDTA

The combination of Tris and EDTA forms the major constituent of lysis buffers. Tris acts as the buffering force, which major role is to maintain the pH of the solution mostly at 8.0. The most common form of Tris, i.e., Tris(hydroxymethyl) aminomethane with p*K*a of 8.1, has its buffering capacity between pH 7 and 9. Tris buffers are generally temperature sensitive and hence should be stored at recommended temperature only. Tris also has an additional property of interacting with lipopolysaccharides (LPS) in the cell membrane, thus destabilizing the cell membrane to aid in cell lysis. Additionally, ethylenediaminetetraacetic acid (EDTA) binds to certain divalent cations mostly calcium (Ca^{2+}) and magnesium (Mg^{2+}). These divalent cations are reported to have their key role in the maintenance of membrane structure and function. Thus, eliminating these ions by EDTA leads to the lysis of cell membrane.

3.3 SDS

Sodium dodecyl sulfate (SDS) is a strong anionic detergent known to have protein denaturation properties. It helps to remove the negative ions from proteins to destabilize their structure and confirmation. In this process, SDS helps in rupturing the cell as well as nuclear membrane to release the genetic material into the solution. Additionally, it helps in releasing DNA from histone and other DNA-binding proteins. SDS is also known to have an additional property of removing lipid membranes, thus aiding in the lysis of cell and nuclear membrane.

3.4 Proteinase-K

Peptide bond present near the carboxyl group of amino acids with blocked alpha amino groups is the cleavage site for the serine protease, i.e., proteinase-K. It helps in degrading many protein impurities to yield a good quality of DNA. It also plays an important role in the inactivation of nucleases, hence preventing the isolated DNA from damage. The activity of proteinase-K is greatly increased by its simultaneous action with SDS and high temperature. Proteinase-K works optimally at 50–65 °C as higher temperature unfolds certain proteins to ease the function of the enzyme.

3.5 Phenol

In most of the cases, phenol is used in Tris-saturated form. Phenol, being a weak acid, is equilibrated with buffers to make it slightly alkaline which is the most suitable pH for DNA isolation. These buffer-saturated phenols contain around 72% of phenol with a slightly higher density than that of water. Phenol has the property to denature proteins during their presence in the aqueous solution. During this process, the nonaqueous component, i.e., protein, gets separated from the aqueous component, i.e., DNA.

3.6 Chloroform-Isoamyl Alcohol

This is mostly used at a ratio of 24:1. These are detergents that bind to the proteins and lipids to dissolve them in the organic solution. After dissolving, the solutions help in clumping the protein-lipid complexes to form a precipitate. Thus, the liquid-liquid extraction technique involving these chemicals distinguishes between the physical positions of the cell components, i.e., upper aqueous phase containing nucleic acids, lower organic phase containing protein components, and middle phase constituting the lipids.

3.7 TE Buffer

TE buffer is generally prepared by mixing 50 mM Tris and 50 mM EDTA at pH 8.0. Tris, the major constituent of TE buffer, acts as a common pH buffer. Additionally, EDTA chelates cations like Mg^{2+}. Hence, TE buffer helps to solubilize DNA by protecting it from degradation.

4 Procedure

1. Take two intact, rooted molar teeth in a 50 ml conical tube containing about 20 ml sterile deionized water.
2. Cap the tubes properly, shake them for 10–12 times, decant the water, and repeat the washing two times.
3. Add 100% ethanol to the tube and repeat the washing as mentioned in step 2 for two times.
4. Clean the teeth with 10% sodium hypochlorite solution followed by 95% ethanol, place the samples on a petri dish, open the lid of the petri dish, and allow the samples to air-dry in a laminar airflow.
5. Take a diamond-cutting disk and cut at the middle line of the tooth, separating the tooth into crown and root.
6. Collect any visible pulp by forceps and place them in a 15 ml conical tube.
7. Crush the root portion of the tooth with the help of a dental drill and place the powder in the same 15 ml conical tube used earlier to collect the pulp.
8. Add 5 ml of forensic buffer to the 15 ml conical tube containing the dental pulp, and monitor whether all the materials dip in the buffer properly.
9. Add proteinase-K to make its final concentration as 100 μg/ml in total volume.
10. Add sodium dodecyl sulfate (SDS) to make its final concentration as 2% in total volume.
11. Cover the cap of the tube in Parafilm®, mix thoroughly, and incubate in water bath at 56 °C for 2–4 h.

12. After 2–4 h, take the samples out of the water bath, add 5 ml (1:1 volume) of Tris-saturated phenol (TSP, pH 8.0), and mix thoroughly for 10 min by gentle shaking.
13. After proper mixing, centrifuge the tube at 10,000 rpm for 10 min at 4 °C.
14. Collect the upper aqueous phase by careful pipetting in a separate conical tube (you may use cut tips).
15. Add half of the volume of TSP (2.5 ml in this case) and chloroform:isoamyl alcohol (24:1) (2.5 ml in this case), and mix thoroughly for 10 min by gentle shaking.
16. After proper mixing, centrifuge the tube at 10,000 rpm for 10 min at 4 °C. Transfer the upper aqueous layer to a fresh conical tube.
17. Add 5 ml (1:1 volume) of chloroform:isoamyl alcohol (24:1), and mix thoroughly for 10 min by gentle shaking.
18. After proper mixing, centrifuge the tube at 10,000 rpm for 10 min at 4 °C.
19. Transfer the supernatant into an Amicon® Ultra-4 ml centrifugal filter.
20. Spin the device at 5000 × *g* for 10 min at 20 °C. Discard the flow through.
21. Add 5 ml of Milli-Q to the device for washing and repeat spinning of the device at 5000 × *g* for 10 min at 20 °C.
22. Recover the concentrated solute by pipetting the solute from the filter and transfer it to a fresh and labeled MCT.
23. Store the DNA at −20 °C for further use.

5 Observation

The quantity of the isolated DNA can be determined at preliminary level by UV–visible spectrophotometer. For a 1 cm path length, the optical density at 260 nm (OD_{260}) equals 1.0 for the following samples:

- 50 μg/ml of double-stranded DNA
- 33 μg/ml of single-stranded DNA
- 20–30 μg/ml of oligonucleotide
- 40 μg/ml of RNA

6 Result Table

Sample	DNA content	$OD_{260/280}$	Inference
Control			
Sample 1 (*molar tooth*)			
Sample 2 (*canine tooth*)			

OD optical density

7 Precautions

- Wear gloves and goggles during manual DNA isolation procedure.
- The plasticware to be used during the entire procedure should be sterilized and DNase free.
- Handle the chemicals carefully, mostly the phenol and chloroform. Phenol, being a strong acid, may cause severe burns whereas chloroform is a known carcinogen.
- Use wide-bore tips by cutting 2–3 mm from the rear end to prevent mechanical disruption of DNA during pipetting.
- Use filter tips to avoid cross-sample contamination.
- Do not use the TSP solution if the color of the solution has changed to pink (occurs due to oxidation of phenol). This may promote nicks in the extracted DNA.
- Pipetting should be done carefully without mixing the two phases, i.e., upper aqueous phase and lower organic phase, at any point of time.
- Clean and sterilize all the physical equipment to be used during mechanical grinding of the tooth samples.
- Choose the tooth sample with intact root for DNA isolation.
- Clean the surface of the teeth carefully prior to mechanical disruption to avoid contamination as well as PCR inhibitors.
- Use appropriate concentration of the reagents (especially sodium hypochlorite) as higher concentration may inhibit further downstream processing of the samples.

8 Troubleshooting

Problem	Tentative cause	Possible solutions
Protein contamination	The extraction step has not been followed properly	Repeat the phenol: chloroform:isoamyl alcohol extraction step
RNA contamination	High-temperature incubation not given to the extracted DNA	Incubate the extracted DNA at 56 °C for 10 min
	RNase not added	Add RNase at a final concentration of 400 μg/ml to the isolated DNA sample
PCR product not amplified	Presence of PCR inhibitors	Clean the surface of tooth samples properly
Partial DNA profile obtained	DNA quantity not sufficient	Select more than one teeth and preferably select molar teeth
Mixed DNA profile obtained	Exogenous contamination	Select intact tooth for DNA isolation
		Clean surface of the samples carefully to remove contaminants
Colored eluent found	Presence of inhibitors (generally occurs in case of teeth recovered from tobacco chewers)	Use alternative DNA isolation protocols such as magnetic bead-based system

Chapter 9

Isolation of DNA from Bone Samples

Objective: To isolate DNA from bone samples.

1 Introduction

Tissues of a human cadaver decompose with time. The factors affecting the decomposition process mostly include various environmental conditions, i.e., climatic conditions, presence of scavengers, and rate of bacterial growth. Over time, the soft tissues tend to degrade; however, tissues of bone remain intact, thus making them a suitable source of DNA for forensic applications. Bone tissues are preferred for DNA-based identification in circumstances such as mutilated bodies, mass disaster, and presence of only skeletal remains in cases of either war, fire explosions, or criminal cases. Though many techniques are used to identify a person from its skeletal remains such as exploring facial and other body characteristics, comparing with other premortem X-rays, superimposition technique, and fingerprint comparison, identification based on DNA typing is the most fallible technique and most widely used in the criminal justice system.

A typical human bone is comprised of either compact or spongy materials due to the presence of connective tissues, protein fibers, and deposition of calcium salts. Osteocytes, the most abundant cells of the bone, cannot divide as they get matured. Osteocytes have two important functions, i.e., maintenance and repair. Other types of cells, the osteoblasts present in the bone matrix, are capable of producing new bone matrix. Similarly, osteoclasts are the large cells having 50 or more number of nuclei which are responsible for recycling of new bone matrix. Hence, the major amount of DNA in found is present in the osteocytes at a magnitude of 20,000–26,000 cells/mm^3. However, degradation and decomposition severely affect the DNA content of the bone samples making them the toughest samples for DNA isolation.

H. R. Dash et al., *Principles and Practices of DNA Analysis: A Laboratory Manual for Forensic DNA Typing*, Springer Protocols Handbooks, https://doi.org/10.1007/978-1-0716-0274-4_9, © Springer Science+Business Media, LLC, part of Springer Nature 2020

Due to the presence of low quality and quantity of DNA samples present in the bone samples they are considered to be challenging from forensic DNA analysis point of view. Femur bone is the preferred sample among the bones of fresh or decomposed cadaver due to longest nature among other human bones. Though any available bone can also be processed, preference should be given to the intact bones. Hence, suitable isolation techniques should be employed to deal with such tough samples. The bone piece should be cleaned properly and carefully prior to the isolation of DNA to avoid environmental inhibitors, microbial contaminations, as well as cross contamination.

2 Principle

The majority of the DNA content of the bone is present inside osteocyte cells. These cells are found embedded in calcified matrix which hinders direct access of the cells during any DNA extraction procedure. Thus, the matrix substances mostly made up of the calcium salts containing collagen fibers should be removed to aid in the final yield of DNA. In this regard, decalcification is the process of removal of calcium salt from tissues such as bone to soften them by the removal of mineralized component. The most common decalcification agents include strong acids, weak acids, and chelating agents. Strong acids such as nitric acid (HNO_3) act as the fastest decalcifier; however, the prolonged exposure to HNO_3 may cause damage to the bone cells. Similarly, formic acid also acts as a decalcifying agent but acts very slowly. The most common decalcifying agent used nowadays in various laboratories is the chelating agent, i.e., EDTA.

Many factors influence the decalcification process of the bone, i.e., the concentration of the decalcifying agent, incubation temperature, and fluid renewal. Use of a larger volume of decalcifying agent expedites the process, whereas incubation at higher temperature also enhances the decalcification process. Additionally, renewal of the decalcifying agent over a period of time increases the rate of removal of calcium ions from bone samples.

Decalcification is followed by the isolation of DNA from the bone cells. In this process, the events unfolded include lysis, purification of DNA from cell debris, and finally concentrating of the isolated DNA. The manual organic DNA isolation process involves the lysis of the cells followed by denaturation and inactivation of proteins using detergents and proteinase-K. Use of phenol, chloroform, and isoamyl alcohol generates the liquid-liquid extraction system which separates nucleic acid from other cell debris by exploiting the differential solubility nature of proteins, lipids, and nucleic acids. Finally, DNA present in the aqueous phase is concentrated and purified by the use of salt and ethanol, followed by its resolubilization in Tris-EDTA buffer.

3 Reagents Required and Their Role

3.1 Sodium Hypochlorite

Sodium hypochlorite (NaOCl) is the most widely used disinfectant and is known for its excellent cleaning action. However, the effectiveness of the solution is dependent on the concentration and pH of the solution. pH of the solution effects the formation of the hypochlorite ion ($^{-}$OCl) and proton (H^+) by dissociation of the weak acid hypochlorous acid (HOCl). In this case, HOCl plays an important role in germicidal action whereas the $^{-}$OCl determines the cleansing efficacy of the solution.

3.2 0.5 mM EDTA

EDTA is a chelating agent employed for the decalcification of bones, the prerequisite for the isolation of DNA from bone samples. The principle behind the functioning of EDTA is that it captures the calcium ions from the surface of crystals, thereby reducing their size. The rate of decalcification of bone using EDTA solution is pH dependent. The action of EDTA is generally occurred at neutral pH; however, the activity of EDTA is enhanced when the pH of the solution is increased to 10. However, this pH adversely affects the bone tissues by damaging them.

3.3 Normal Saline Solution (NSS)

0.9% Solution of sodium chloride is called as normal saline. It is mostly used for washing of living cells to remove other debris. As this concentration of salt solution is isotonic to the cells, it does not cause lysis or disruption of the cells. However, prior to using the prepared solution should be sterilized by either autoclaving or passing through a 0.45 or 0.22 μm sterile filter.

3.4 Tris-EDTA

The combination of Tris and EDTA forms the major constituent of lysis buffers. Tris acts as the buffering force, whose major role is to maintain pH of the solution mostly at 8.0. The most common form of Tris, i.e., Tris(hydroxymethyl) aminomethane with p*K*a of 8.1, has its buffering capacity between pH 7 and 9. Tris buffers are generally temperature sensitive and hence should be stored at recommended temperature only. Tris also has an additional property of interacting with lipopolysaccharides (LPS) in the cell membrane, thus destabilizing the cell membrane to aid in cell lysis. Additionally, ethylenediaminetetraacetic acid (EDTA) binds to certain divalent cations mostly calcium (Ca^{2+}) and magnesium (Mg^{2+}). These divalent cations are reported to have their key role in the maintenance of membrane structure and function. Thus, eliminating these ions by EDTA leads to the lysis of cell membrane.

3.5 SDS

Sodium dodecyl sulfate (SDS) is a strong anionic detergent known to have protein denaturation properties. It helps to remove the negative ions from proteins to destabilize their structure and confirmation. In this process, SDS helps in rupturing the cell as well as nuclear membrane to release the genetic material into the solution. Additionally, it helps in releasing DNA from histone and other DNA-binding proteins. SDS is also known to have an additional property of removing lipid membranes, thus aiding in the lysis of cell and nuclear membrane.

3.6 Proteinase-K

Peptide bond present near the carboxyl group of amino acids with blocked alpha amino groups is the cleavage site for the serine protease, i.e., proteinase-K. It helps in degrading many protein impurities to yield a good quality of DNA. It also plays an important role in the inactivation of nucleases, hence preventing the isolated DNA from damage. The activity of proteinase-K is greatly increased by its simultaneous action with SDS and high temperature. Proteinase-K works optimally at 50–65 °C as higher temperature unfolds certain proteins to ease the function of the enzyme.

3.7 DTT

Dithiothreitol (DTT) prevents the formation of intramolecular and intermolecular disulfide bonds among cysteine residues. With protein disulfide bridges being the major constituent of sperm nuclear membrane, DTT is essential during lysis of sperm cells to release sperm DNA. Hence, during the lysis of sperm cells, in addition to proteinase-K and SDS, DTT is also used in the mixture for their effective lysis.

3.8 Phenol

In most of the cases, phenol is used in Tris-saturated form. Phenol, being a weak acid, is equilibrated with buffers to make it slightly alkaline which is the most suitable pH for DNA isolation. These buffer-saturated phenols contain around 72% of phenol with a slightly higher density than that of water. Phenol has the property to denature proteins during their presence in the aqueous solution. During this process, the nonaqueous component, i.e., protein, gets separated from the aqueous component, i.e., DNA.

3.9 Chloroform-Isoamyl Alcohol

This is mostly used at a ratio of 24:1. These are detergents that bind to the proteins and lipids to dissolve them in the organic solution. After dissolving, the solutions help in clumping the protein-lipid complexes to form a precipitate. Thus, the liquid-liquid extraction technique involving these chemicals distinguishes between the physical positions of the cell components, i.e., upper aqueous phase containing nucleic acids, lower organic phase containing protein components, and middle phase constituting the lipids.

3.10 TE Buffer

TE buffer is generally prepared by mixing 50 mM Tris and 50 mM EDTA at pH 8.0. Tris, the major constituent of TE buffer, acts as a common pH buffer. Additionally, EDTA chelates cations like Mg^{2+}. Hence, TE buffer helps to solubilize DNA by protecting it from degradation.

4 Procedure

1. Select intact long bones, preferably femur, humerus, radius, ulna, or skull.
2. Immerse the bone(s) in 3% sodium hypochlorite (NaOCl) (prepared in molecular biology-grade water) for 15 min.
3. Optionally clean the surface of the bone using sandpaper for old, ancient, or excavated bones.
4. Place the bone on a fresh aluminum foil.
5. Drill holes in the bone using a mechanical drilling machine while another person holding the bone sample tightly.
6. Collect the bone powder in a sterile aluminum foil and transfer it to a sterile 50 ml centrifuge tube.
7. Add 0.5 M EDTA pH 8.0 sufficient enough to immerse the bone powder.
8. Incubate the tube containing bone powder in 0.5 M EDTA solution at 37 °C for 72 h with repeated shaking.
9. After 72 h, centrifuge the tube at 10,000 rpm for 15 min at 15 °C.
10. Discard the supernatant and wash the bone powder with the same amount of sterile saline solution thoroughly with vortex; centrifuge the tube at 10,000 rpm for 15 min at 15 °C.
11. Repeat the washing step three times for better removal of debris.
12. Add forensic buffer (composition given in Appendix A) in appropriate quantity sufficient enough to immerse the bone powder.
13. Add proteinase K and SDS to a final concentration of 200 μg/ml and 2%, respectively.
14. Add DTT to make its final concentration as 100 μg/ml in total volume.
15. Cover the cap of the tube in Parafilm®, mix thoroughly, and incubate in water bath at 56 °C for 2 h to overnight.
16. Take the samples out of the water bath, and decant the lysate to a fresh tube.
17. Add 1:1 volume of Tris-saturated phenol (TSP, pH 8.0) to the lysate, and mix thoroughly for 10 min by gentle shaking.
18. After proper mixing, centrifuge the tube at 10,000 rpm for 10 min at 4 °C.
19. Collect the upper aqueous phase by careful pipetting in a separate conical tube (you may use cut tips).

20. Add half of the volume of TSP and chloroform:isoamyl alcohol (24:1), and mix thoroughly for 10 min by gentle shaking.
21. After proper mixing, centrifuge the tube at 10,000 rpm for 10 min at 4 °C. Transfer the upper aqueous layer to a fresh conical tube.
22. Add 1:1 volume of chloroform:isoamyl alcohol (24:1), and mix thoroughly for 10 min by gentle shaking.
23. After proper mixing, centrifuge the tube at 10,000 rpm for 10 min at 4 °C.
24. Transfer the supernatant into an Amicon® Ultra-4 ml centrifugal filter.
25. Spin the device at 5000 × *g* for 10 min at 20 °C. Discard the flow through.
26. Add 5 ml of Milli-Q to the device for washing and repeat spinning of the device at 5000 × *g* for 10 min at 20 °C.
27. Recover the concentrated solute by pipetting the solute from the filter and transfer it to a fresh and labeled MCT.
28. Store the DNA at −20 °C for further use.

5 Observation

The quantity of the isolated DNA can be determined at preliminary level by UV–visible spectrophotometer. For a 1 cm path length, the optical density at 260 nm (OD_{260}) equals 1.0 for the following samples:

- 50 μg/ml of double-stranded DNA
- 33 μg/ml of single-stranded DNA
- 20–30 μg/ml of oligonucleotide
- 40 μg/ml of RNA

6 Result Table

Sample	DNA content	$OD_{260/280}$	Inference
Control			
Sample 1 (*femur*)			
Sample 2 (*skull*)			

OD optical density

7 Precautions

- Wear gloves and goggles during manual DNA isolation procedure.
- The plasticware to be used during the entire procedure should be sterilized and DNase free.
- Handle the chemicals carefully, mostly the phenol and chloroform. Phenol, being a strong acid, may cause severe burns whereas chloroform is a known carcinogen.
- Use wide-bore tips by cutting 2–3 mm from the rear end to prevent mechanical disruption of DNA during pipetting.
- Use filter tips to avoid cross-sample contamination.
- Do not use the TSP solution if the color of the solution has changed to pink (occurs due to oxidation of phenol); this may promote nicks in the extracted DNA.
- Pipetting should be done carefully without mixing the two phases, i.e., upper aqueous phase and lower organic phase, at any point of time.
- Choose intact piece of bone sample (if available) for DNA isolation.
- Use appropriate concentration of the reagents (especially sodium hypochlorite) as higher concentration may inhibit the further downstream processing of the samples.

8 Troubleshooting

Problem	Tentative cause	Possible solutions
Protein contamination	The extraction step has not been followed properly	Repeat the phenol: chloroform:isoamyl alcohol extraction step
RNA contamination	High-temperature incubation not given to the extracted DNA	Incubate the extracted DNA at 56 °C for 10 min
	RNase not added	Add RNase at a final concentration of 400 μg/ml to the isolated DNA sample
PCR product not amplified	Presence of PCR inhibitors	Clean the bone surface properly
Partial DNA profile obtained	DNA quantity not sufficient	Increase the sample size; take more amount of bone powder

(continued)

Problem	Tentative cause	Possible solutions
Mixed DNA profile obtained	Exogenous contamination	Select intact bone for DNA isolation
		Surface clean the samples carefully to remove contaminants
Colored eluent found	Presence of inhibitors (generally occurs in old/excavated bones)	Use alternative DNA isolation protocols such as magnetic bead-based system
		Clean the surface of old bones properly by sandpaper prior to processing

Chapter 10

Differential Extraction of Sperm and Epithelial Cell DNA

Objective: To isolate sperm and epithelial cell DNA differentially from mixed samples (vaginal fluid slide) of sexual assault victims.

1 Introduction

Sexual assault has become a huge problem globally. Many reasons determine the fact that a victim could not identify an offender. In this process, DNA-based identification is left with the only alternative to identify the perpetrator. In sexual assault cases, it is of utmost importance to establish the presence of victim's DNA as well as the identification of perpetrator from the examined objects. Over the years, DNA technology has emerged to be the most useful tool for identification with high discrimination potential and sensitivity. In sexual assault cases, the most common sample for examination is victim's vaginal fluid slide. The glass slide containing sample solves two purposes, i.e., isolation of DNA and screening of sperm cells via optical microscopy. The major challenge in processing such samples is to generate a mixed DNA profile of both victim and perpetrator. Thus, many a times, a unique DNA extraction step is followed to simultaneously isolate male and female DNA fragments from the same sample in a differential manner.

Perpetrator's DNA is the most likely to be obtained from the sperm cells present on the victim's sources routinely collected during medical examination. However, the majority fraction of the DNA present on the victim's belongings is of female/victim origin constituting the epithelial cells of the vaginal lining. These epithelial cells should be separated from the sperm cells prior to the isolation of DNA from the sperm cells using the differential DNA isolation procedure described in Fig. 10.1.

H. R. Dash et al., *Principles and Practices of DNA Analysis: A Laboratory Manual for Forensic DNA Typing*, Springer Protocols Handbooks, https://doi.org/10.1007/978-1-0716-0274-4_10, © Springer Science+Business Media, LLC, part of Springer Nature 2020

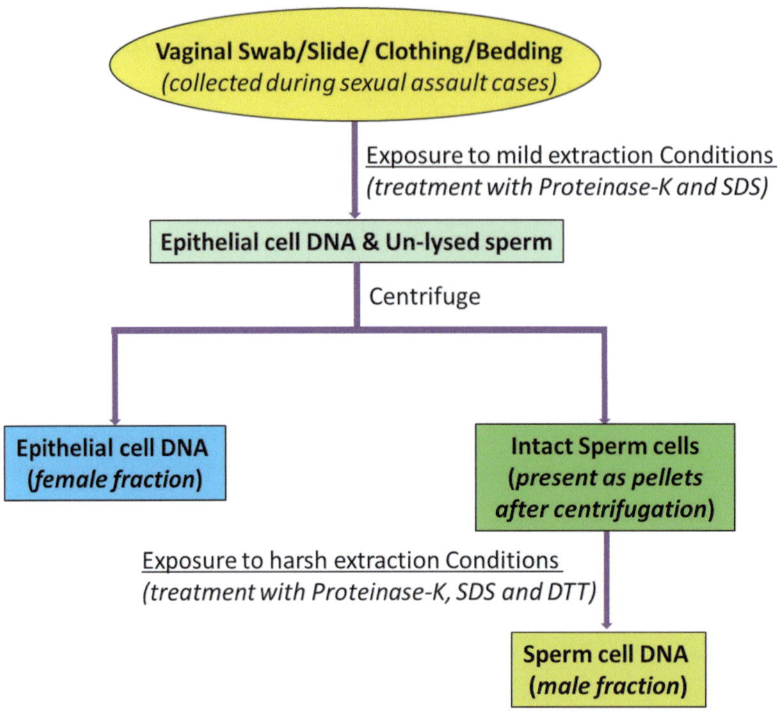

Fig. 10.1 Principle of differential extraction technique to separate male and female DNA fractions from mixed samples

2 Principle

The most common evidences collected in sexual assault cases consist of swabs of vagina; oral, rectal, and exposed skin; as well as clothing and bedding from the spot. These evidences contain stains constituting both victim's cells and seminal stains containing spermatozoa from the accused. The most common technique for processing of these samples involve isolation of DNA based on two separate cellular fractions, i.e., non-sperm, epithelial cell fractions of female origin and sperm cell fraction originating from male or perpetrator.

The first step of DNA isolation involves the isolation of DNA from the non-sperm cells by using chemical lysis technique involving buffers, proteinase-K, and SDS. During this step, the non-sperm epithelial cells are digested to release their DNA to the solution and the sperm heads remain intact in this process. Further, the sperm heads are separated from epithelial DNA and other cellular debris by centrifugation and repeated washing. In the second step of DNA isolation, the sperm heads are subjected to digestion using proteinase-K, SDS, and dithiothreitol (DTT). DTT, the reducing agent, disrupts the disulfide bond of the structural proteins present on the sperm cells.

The differential steps involving DNA isolation can be followed only when there is the presence of sperm cells in the sample which needs to be confirmed prior to DNA isolation. The preliminary steps involving the confirmation of the presence of spermatozoa are either microscopic or immunological tests. The most common immunologic test for the presence of semen involves the detection of a protein P30 alternatively called as prostate-specific antigen. Additionally, staining of the cells by Christmas tree stain followed by observation under microscope can also confirm the presence of spermatozoa in the samples. Differential DNA isolation technique should be performed only after confirming the presence of spermatozoa in the target samples.

3 Reagents Required and Their Role

3.1 Tris-EDTA

The combination of Tris and EDTA forms the major constituent of lysis buffers. Tris acts as the buffering force, whose major role is to maintain pH of the solution mostly at 8.0. The most common form of Tris, i.e., Tris(hydroxymethyl) aminomethane with p*Ka* of 8.1, has its buffering capacity between pH 7 and 9. Tris buffers are generally temperature sensitive and hence should be stored at recommended temperature only. Tris also has an additional property of interacting with lipopolysaccharides (LPS) in the cell membrane, thus destabilizing the cell membrane to aid in cell lysis. Additionally, ethylenediaminetetraacetic acid (EDTA) binds to certain divalent cations mostly calcium (Ca^{2+}) and magnesium (Mg^{2+}). These divalent cations are reported to have their key role in the maintenance of membrane structure and function. Thus, eliminating these ions by EDTA leads to the lysis of cell membrane.

3.2 SDS

Sodium dodecyl sulfate (SDS) is a strong anionic detergent known to have protein denaturation properties. It helps to remove the negative ions from proteins to destabilize their structure and confirmation. In this process, SDS helps in rupturing the cell as well as nuclear membrane to release the genetic material into the solution. Additionally, it helps in releasing DNA from histone and other DNA-binding proteins. SDS is also known to have an additional property of removing lipid membranes, thus aiding in the lysis of cell and nuclear membrane.

3.3 Proteinase-K

Peptide bond present near the carboxyl group of amino acids with blocked alpha amino groups is the cleavage site for the serine protease, i.e., proteinase-K. It helps in degrading many protein impurities to yield a good quality of DNA. It also plays an important role in the inactivation of nucleases, hence preventing the isolated DNA from damage. The activity of proteinase-K is greatly increased by its simultaneous action with SDS and high temperature. Proteinase-K works optimally at 50–65 °C as higher temperature unfolds certain proteins to ease the function of the enzyme.

3.4 DTT

Dithiothreitol (DTT) prevents the formation of intramolecular and intermolecular disulfide bonds among cysteine residues. With protein disulfide bridges being the major constituent of sperm nuclear membrane, DTT is essential during lysis of sperm cells to release sperm DNA. Hence, during the lysis of sperm cells, in addition to proteinase-K and SDS, DTT is also used in the mixture for their effective lysis.

3.5 Phenol

In most of the cases, phenol is used in Tris-saturated form. Phenol, being a weak acid, is equilibrated with buffers to make it slightly alkaline which is the most suitable pH for DNA isolation. These buffer-saturated phenols contain around 72% of phenol with a slightly higher density than that of water. Phenol has the property to denature proteins during their presence in the aqueous solution. During this process, the nonaqueous component, i.e., protein, gets separated from the aqueous component, i.e., DNA.

3.6 Chloroform-Isoamyl Alcohol

This is mostly used at a ratio of 24:1. These are detergents that bind to the proteins and lipids to dissolve them in the organic solution. After dissolving, the solutions help in clumping the protein-lipid complexes to form a precipitate. Thus, the liquid-liquid extraction technique involving these chemicals distinguishes between the physical positions of the cell components, i.e., upper aqueous phase containing nucleic acids, lower organic phase containing protein components, and middle phase constituting the lipids.

3.7 TE Buffer

TE buffer is generally prepared by mixing 50 mM Tris and 50 mM EDTA at pH 8.0. Tris, the major constituent of TE buffer, acts as a common pH buffer. Additionally, EDTA chelates cations like Mg^{2+}. Hence, TE buffer helps to solubilize DNA by protecting it from degradation.

4 Procedure

1. Take a properly labeled sterile microcentrifuge tube (MCT) containing 500 μl of forensic buffer.
2. Transfer the cellular material present in the vaginal slide by the help of a cell scrapper using this forensic buffer to the previously marked tube.
3. Repeat the process 2–3 times so that significant number of cells are present in the buffer (the buffer in the MCT will turn translucent from transparent when the cellular materials of the slide come into the solution; check this).

4. Add proteinase-K to make its final concentration as 100 μg/ml in total volume.
5. Add sodium dodecyl sulfate (SDS) to make its final concentration as 2% in total volume.
6. Cover the cap of the tube in Parafilm®, mix thoroughly, and incubate in water bath at 56 °C for 2–4 h.
7. Centrifuge the digested material of the sample for 5 min at 12,000 × *g* at 15 °C.
8. Transfer the liquid portion of the solution into a fresh tube (*tube I*) (this solution contains the digested non-sperm cells).
   (a) Add 500 μl of Tris-saturated phenol (TSP, pH 8.0) to *tube I*, and mix properly for 10 min by gentle shaking.
   (b) After proper mixing, centrifuge the tube at 10,000 rpm for 10 min at 4 °C.
   (c) Collect the upper aqueous phase by careful pipetting in a separate MCT (you may use cut tips).
   (d) Add half of the volume of TSP (250 μl in this case) and chloroform:isoamyl alcohol (24:1) (250 μl in this case), and mix thoroughly for 10 min by gentle shaking.
   (e) After proper mixing, centrifuge the tube at 10,000 rpm for 10 min at 4 °C. Transfer the upper aqueous layer to a fresh MCT.
   (f) Add 500 μl (1:1 volume) of chloroform:isoamyl alcohol (24:1), and mix thoroughly for 10 min by gentle shaking.
   (g) After proper mixing, centrifuge the tube at 10,000 rpm for 10 min at 4 °C.
   (h) Transfer the supernatant into an Amicon® Ultra-0.5 ml 30 kDa centrifugal filter placed in a 2.0 ml MCT.
   (i) Spin the device at 5000 × *g* for 10 min at 20 °C. Discard the flow through.
   (j) Add 500 μl of Milli-Q to the device for washing and repeat spinning of the device at 5000 × *g* for 10 min at 20 °C.
   (k) Remove the assembled device and to recover the concentrated solute place the Amicon® ultra-0.5 ml 30 kDa centrifugal filter in a fresh MCT upside down.
   (l) Counterbalance with a similar device and spin at 1000 × *g* for 2 min at 20 °C.
   (m) Recover the concentrated solute by pipetting the solute from the filter and transfer it to a fresh and labeled MCT (*female fraction*).
   (n) Store the DNA at −20 °C for further use.

9. Wash the pellet twice with pre-sterilized normal saline solution.
10. Add 300 μl of forensic buffer to the pellet obtained after centrifugation and separation of liquid portion from it (*tube II*) (the pellet contains the sperm cells or the male fraction).
11. To *tube II*, add proteinase-K to make its final concentration as 100 μg/ml in total volume.
12. To *tube II*, add sodium dodecyl sulfate (SDS) to make its final concentration as 2% in total volume.
13. To *tube II*, add dithiothreitol (DTT) to make its final concentration as 150 mM in total volume.
14. Cover the cap of the tube with Parafilm®, mix thoroughly, and incubate in water bath at 50 °C for 2–4 h.
    (a) After 2–4 h, take the samples out of the water bath, add 300 μl (1:1 volume) of Tris-saturated phenol (TSP, pH 8.0), and mix thoroughly for 10 min by gentle shaking.
    (b) After proper mixing, centrifuge the tube at 10,000 rpm for 10 min at 4 °C.
    (c) Collect the upper aqueous phase by careful pipetting in a separate conical tube (you may use cut tips).
    (d) Add half of the volume of TSP (150 μl in this case) and chloroform:isoamyl alcohol (24:1) (150 μl in this case), and mix thoroughly for 10 min by gentle shaking.
    (e) After proper mixing, centrifuge the tube at 10,000 rpm for 10 min at 4 °C. Transfer the upper aqueous layer to a fresh MCT.
    (f) Add 300 μl (1:1 volume) of chloroform:isoamyl alcohol (24:1), and mix thoroughly for 10 min by gentle shaking.
    (g) After proper mixing, centrifuge the tube at 10,000 rpm for 10 min at 4 °C.
    (h) Transfer the supernatant into an Amicon® Ultra-0.5 ml 30 kDa centrifugal filter placed in a 2.0 ml MCT.
    (i) Spin the device at 5000 × *g* for 10 min at 20 °C. Discard the flow through.
    (j) Add 450 μl of Milli-Q to the device for washing and repeat spinning of the device at 5000 × *g* for 10 min at 20 °C.
    (k) Recover the concentrated solute by pipetting the solute from the filter and transfer it to a fresh and labeled MCT (*male fraction*).
    (l) Store the DNA at −20 °C for further use.

5 Observation

The quantity of the isolated DNA can be determined at preliminary level by UV–visible spectrophotometer. For a 1 cm path length, the optical density at 260 nm (OD_{260}) equals 1.0 for the following samples:

- 50 μg/ml of double-stranded DNA
- 33 μg/ml of single-stranded DNA
- 20–30 μg/ml of oligonucleotide
- 40 μg/ml of RNA

6 Result Table

Sample	DNA content	$OD_{260/280}$	Inference
Control			
Sample 1 (*female fraction*)			
Sample 2 (*male fraction*)			

OD optical density

7 Precautions

- Wear gloves and goggles during manual DNA isolation procedure.
- The plasticware to be used during the entire procedure should be sterilized and DNase free.
- Handle the chemicals carefully, mostly the phenol and chloroform. Phenol, being a strong acid, may cause severe burns whereas chloroform is a known carcinogen.
- Use wide-bore tips by cutting 2–3 mm from the rear end to prevent mechanical disruption of DNA during pipetting.
- Use filter tips to avoid cross-sample contamination.
- Do not use the TSP solution if the color of the solution has changed to pink (occurs due to oxidation of phenol). This may promote nicks in the extracted DNA.
- Pipetting should be carried out carefully without mixing the two phases, i.e., upper aqueous phase and lower organic phase, at any point of time.
- Prior to DNA isolation confirm the presence of spermatozoa in the vaginal slide by microscopic techniques.

8 Troubleshooting

Problem	Tentative cause	Possible solutions
Protein contamination	The extraction step has not been followed properly	Repeat the phenol: chloroform:isoamyl alcohol extraction step
RNA contamination	High-temperature incubation not given to the extracted DNA	Incubate the extracted DNA at 56 °C for 10 min
	RNase not added	Add RNase at a final concentration of 400 μg/ml to the isolated DNA sample
PCR product not amplified for male DNA	Spermatozoa not present	Confirm the presence of spermatozoa prior to DNA isolation
Partial DNA profile obtained	DNA quantity not sufficient	Quantify the sample prior to PCR amplification and take optimum quantity of DNA
Mixed DNA profile obtained from male fraction	Presence of female DNA in the sample in addition to male DNA	The alleles with higher peak height are from male contributor, whereas the alleles with lower peak height are from female contributor
Mixed DNA profile obtained from female fraction	Presence of male DNA in the sample in addition to female DNA	The alleles with higher peak height are from female contributor, whereas the alleles with lower peak height are from male contributor

Part III

Automatic and Semi-automatic Extraction of DNA from Biological Samples

Chapter 11

Isolation of DNA by Using Magnetic Bead-Based Extraction System

Objective: To isolate DNA from liquid blood and forensic samples using magnetic bead-based extraction system.

1 Introduction

Isolation of DNA is the utmost important pre-requirement and fundamental for forensic analysis and criminal justice system by DNA fingerprinting technique. Since the first extraction of DNA in 1869 by Friedrich Miescher, many techniques have been discovered so far for effective extraction of DNA from biological samples. Pioneer extraction was a simple discovery which relied on the fact that inside the cell there exists some materials which precipitate out of the acidic solution whereas they dissolve in alkaline solution. However the material DNA was discovered in 1953.

Traditional method of DNA extraction involves a liquid-liquid extraction procedure consisting of three basic steps, i.e., lysis of cell, removal of other cell debris from nucleic acids, and precipitation of nucleic acids. In this technique physical/chemical disruption of cells takes place followed by the use of phenol-chloroform organic solvents to concentrate DNA in hydrophilic phase. Finally, DNA pellet is collected at the bottom of the solution followed by its dissolving in water. The methodologies of manual extraction of DNA involve manyfold manipulation of the samples as well as the reagents increasing the chance of contamination. Another shortcoming of manual extraction of DNA involves the consumption of time for the process and requirement of individual's attention and labor till completion of the process. Additionally, the yield of DNA by manual extraction techniques is always dependent on the chemicals used and the examiner's proficiency. In this regard, with the increase in goal to streamline the process of DNA extraction

H. R. Dash et al., *Principles and Practices of DNA Analysis: A Laboratory Manual for Forensic DNA Typing*, Springer Protocols Handbooks, https://doi.org/10.1007/978-1-0716-0274-4_11, © Springer Science+Business Media, LLC, part of Springer Nature 2020

throughout the globe, other solid-phase-based systems are being used for isolation and purification of DNA. Magnetic bead-based DNA isolation technique has been discussed in much details in this chapter.

2 Principle

The first report of using magnetic particles for extraction of DNA was in the year 1998. The initial technique of magnetic separation involves the use of particles constituting iron oxide core coated with silane. The surface of these particles is bound with free carboxylic acid molecules which in turn are bound to the nucleic acids. In this case, the concentration of the salts determines the strength of controlled reversible bonding between the DNA/RNA and the functional groups. Thus, at optimum concentration of the salts, the magnetic particles are attached to the nucleic acids. Finally, a magnetic field is created by placing a magnet outside the tube, enabling the magnetic particles containing nucleic acids to stick to the outer edge of the tube. This step is followed by the washing of the particles with washing buffer to release the nucleic acid from the magnetic beads to generate a pure nucleic acid extract. Figure 11.1 describes the schematic representation of the principle of DNA extraction by magnetic bead-based technique.

There are many commercially available kits in the market for magnetic bead-based extraction of DNA from forensic samples. Two magnetic bead-based kits with potential application of DNA isolation from a wider variety of samples have been included in this chapter, i.e., PrepFiler Express™ and PrepFiler Express BTA™ kits

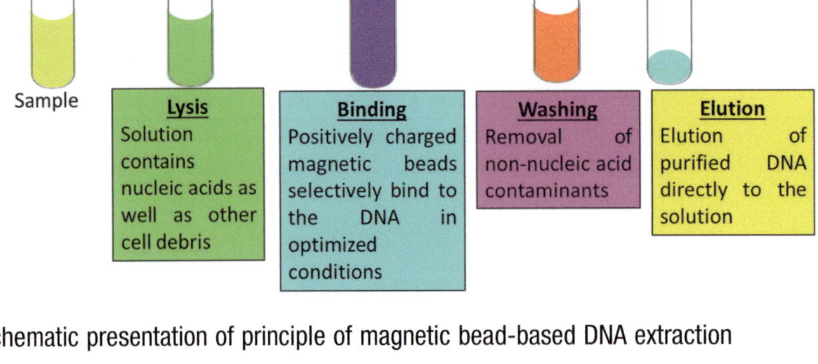

Fig. 11.1 Schematic presentation of principle of magnetic bead-based DNA extraction

(Applied Biosystems) for extraction of DNA from several forensic samples including blood, saliva, body fluid stains, hair root, cigarette butt, chewing gum, bone, and tooth in an efficient manner. The other available kits for isolation of DNA from forensic samples employing magnetic bead-based principle have been listed in Appendix B.

3 Reagents Required and Their Role

3.1 PrepFiler™ Lysis Buffer

The buffer is provided by the manufacturer. This buffer is used for the lysis of common forensic samples including liquid blood, liquid saliva, bloodstains, other body fluid stains, body fluids on swabs, and hair root.

3.2 PrepFiler BTA™ Lysis Buffer

The buffer is also provided by the manufacturer. This buffer is used for the lysis of challenging samples such as chewing gum, cigarette butt, and tape lifts as well as other tough samples, i.e., bone and tooth.

3.3 LySep™ Column

LySep™ column provides a useful tool for simultaneous lysis and purification of the lysate in the same tube. It consists of two parts, the LySep™ column and the sample tube, and the insertion of the column to the tube forms the LySep™ column assembly (Fig. 11.2). The sample as well as the lysis buffer are added to the LySep™ column assembly sequentially followed by lysis in a thermo-shaker/mixer. After completion of lysis process, the assembly is centrifuged to deform the LySep™ column allowing the lysate to pass through the column to the sample tube. The LySep™ column and the substrate can be discarded and the sample tube containing sample can be applied to the instrument for further processing.

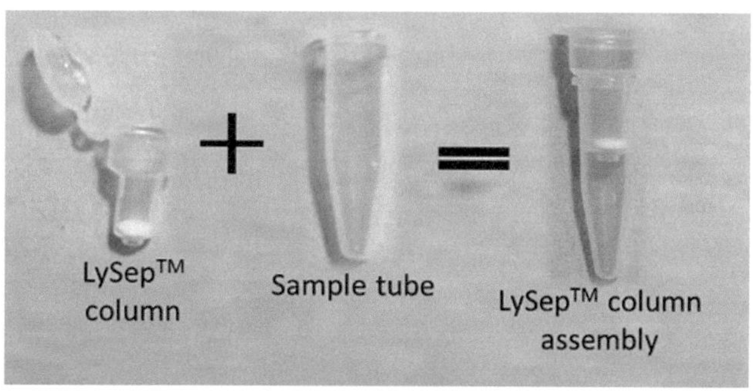

Fig. 11.2 Assembling of LySep™ column for lysis of forensic sample

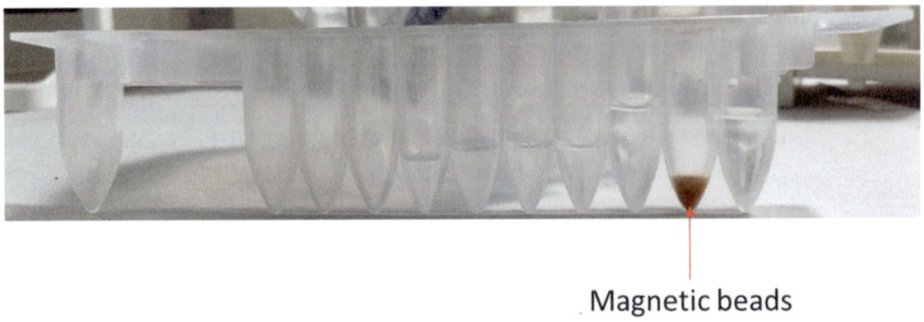

Fig. 11.3 Cartridge containing reagents and magnetic beads for purification of DNA from cell lysate

3.4 Cartridges

All the reagents and chemicals required for purification of DNA from lysed cell debris are stored in the sealed plastic containers called as cartridges (Fig. 11.3). Mostly, the cartridges contain lysis buffer, binding solution, elution buffer, washing buffer, and magnetic beads in suspension. Additionally, some compartments are left empty to provide space for heated chamber for elution. The cartridges are generally pre-sealed with aluminum sheets and stored at room temperature. Prior to the automated isolation steps by the instrument, automated piercing steps are followed which minimizes possible contamination and buffer spout. Finally, mixing of the beads, reagents, and sample is carried out by the liquid-handling action of the instrument. Additionally, separation of DNA is achieved by magnetic attraction as well as liquid handling. The cartridges are specific to the isolation kits and the instrument used.

3.5 Proteinase-K

Peptide bond present near the carboxyl group of amino acids with blocked alpha amino groups is the cleavage site for the serine protease, i.e., proteinase-K. It helps in degrading many protein impurities to yield a good quality of DNA. It also plays an important role in the inactivation of nucleases, hence preventing the isolated DNA from damage. The activity of proteinase-K is greatly increased by its simultaneous action with SDS and high temperature. Proteinase-K works optimally at 50–65 °C as higher temperature unfolds certain proteins to ease the function of the enzyme.

3.6 DTT

Dithiothreitol (DTT) prevents the formation of intramolecular and intermolecular disulfide bonds among cysteine residues. With protein disulfide bridges being the major constituent of sperm nuclear membrane, DTT is essential during lysis of sperm cells to release sperm DNA. Hence, during the lysis of sperm cells, in addition to proteinase-K and SDS, DTT is also used in the mixture for their effective lysis.

4 Procedure

4.1 Using PrepFiler Express™ Forensic DNA Extraction Kit

1. Check PrepFiler Express™ lysis buffer for the presence of suspected precipitate. If suspected precipitate matter is found, heat the solution at 37 °C followed by vortex for 5 s.
2. Prepare the lysis solution freshly by mixing 500 μl lysis buffer with 5 μl freshly prepared 1 M DTT.
3. Insert a LySep™ column into a sample tube followed by careful transfer of the sample into LySep™ column.
4. Add 500 μl freshly prepared lysis solution to the column assembly containing the sample. Make sure that the entire sample is submerged in the lysis solution.

   Note: For optimum extraction of DNA from the forensic samples, input appropriate amount of sample, i.e., liquid blood/saliva (40 μl), blood-stained clothing (25 mm^2 cutting/punch), body fluid stains on clothing (25 mm^2 cutting), and hair root (5 mm cutting from the hair root).
5. Carefully close the lid of the assembled column.
6. Incubate the column containing samples in a thermal shaker at 70 °C and 750 rpm for 40 min.
7. After incubation, centrifuge the column assembly for 2 min at 10,000 × *g* for transfer of lysate to the sample tube.
8. Carefully remove the LySep™ column from the sample tube and discard the LySep™ column.
9. Proceed the samples directly to automated extraction system.

4.2 Using PrepFiler Express BTA™ Forensic DNA Extraction Kit

1. Prepare a fresh lysis solution containing 220 μl lysis buffer, 3 μl freshly prepared 1 M DTT, and 7 μl proteinase-K.
2. Place the sample (bone/teeth) in a PrepFiler™ Bone and tooth lysate tube.
3. Add 230 μl of freshly prepared lysis solution to the lysate tube containing bone/teeth samples.

   Note: For optimum extraction of DNA from bone and teeth samples, input appropriate amount of samples, i.e., up to 50 g of bone and teeth powder.
4. Cover the lysate tubes containing the samples and the lysis solution properly and vortex gently for 5 s.
5. Incubate the lysate tubes containing samples in a thermal shaker at 56 °C and 1100 rpm for 2–3 h.
6. Centrifuge the tubes at 10,000 × *g* for 91 s at 4 °C.
7. Transfer the clear lysate to a fresh PrepFiler™ sample tube.
8. Proceed the samples directly for automated extraction system.

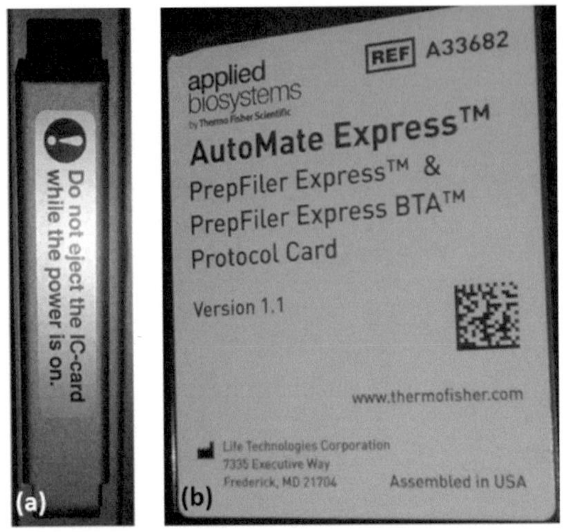

Fig. 11.4 (**a**) Card slot of the instrument and (**b**) protocol card

4.3 Setup and Run for Automated DNA Extraction

1. Prior to switching on the instrument, insert the appropriate protocol card (either PrepFiler Express™ or PrepFiler Express BTA™) in the card slot, and close the card slot. For guidance see Fig. 11.4.
2. Switch on the instrument; the display of the instrument shows information regarding the instrument version and the main menu; press "Start." For guidance see Fig. 11.5.
3. Press ⏎ after following each prompt on-screen.
4. Open the instrument door to remove the cartridge rack and tip and tube rack from the instrument.
5. Load the desired numbers of prefilled reagent cartridges (depending on the number of samples) into the cartridge rack. Insert the cartridge rack in the instrument. For guidance see Fig. 11.6.
6. Load the tip and tube rack. For guidance see Fig. 11.7.
7. Place the PrepFiler™ sample tubes containing lysate samples in Row S corresponding to the position of cartridges.
8. Place the AutoMate Express™ tips inserted in tip holders in Row T2 corresponding to the position of sample tubes.
9. Place the labeled PrepFiler™ elution tubes in open condition in Row E corresponding to the position of sample tubes.
10. Position T1 should be left empty.
11. Insert the loaded tip and tube rack into the instrument.
12. Close the instrument door properly.
13. Press ⏎ after closing the door.

Fig. 11.5 "Display" of the instrument showing information regarding the instrument and main menu

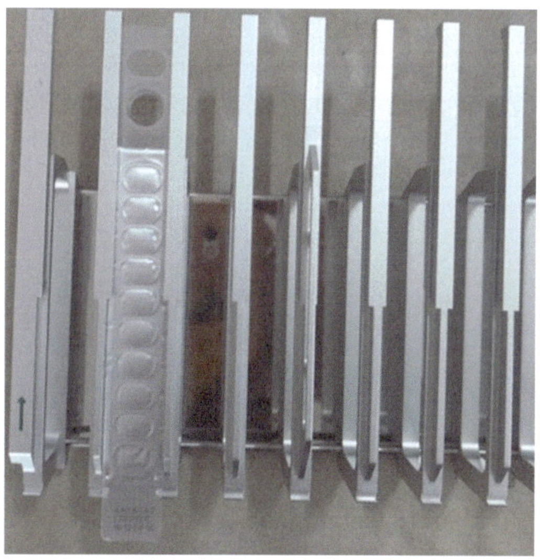

Fig. 11.6 Loading of the prefilled reagent cartridges into the cartridge rack of the instrument

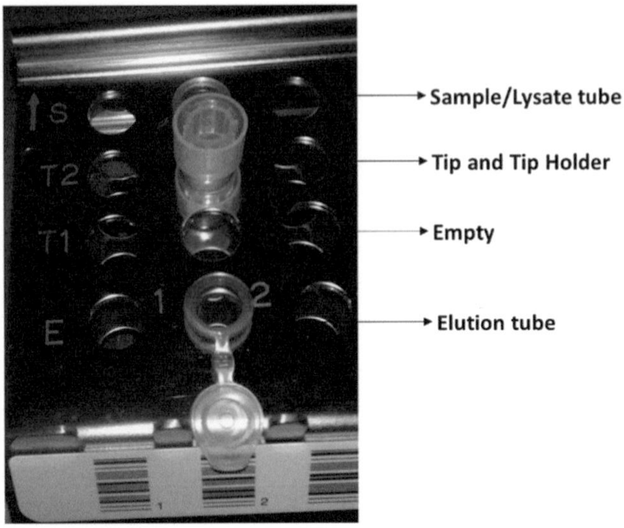

Fig. 11.7 Correct position of the sample tubes, elution tubes, and tips and tip holders on the tip and tube rack

14. Press 1, if you want to use protocol for PrepFiler Express™ kit; else press 2 for the use of PrepFiler Express BTA™ kit.
15. Press "Start"; the screen shows the real-time steps of the extraction procedure and the remaining time of the extraction process.
16. After completion of the run, press ⮐ to return to the main menu.
17. Open the instrument door, and remove the cartridge rack as well as the tip and tube rack.
18. Remove the elution tubes containing the purified DNA and they can be stored at −20 °C till further use.
19. Dispose the exhausted reagent cartridges, tips, and sample tubes properly.

5 Observation

The quantity of the isolated DNA can be determined at the preliminary level by UV–visible spectrophotometer. For a 1 cm path length, the optical density at 260 nm (OD_{260}) equals 1.0 for the following samples:

- 50 μg/ml of double-stranded DNA
- 33 μg/ml of single-stranded DNA
- 20–30 μg/ml of oligonucleotide
- 40 μg/ml of RNA

6 Result Table

Sample	DNA content	$OD_{260/280}$	Inference
Control			
Sample 1 (*liquid blood*)			
Sample 2 (*bone powder*)			

OD optical density

7 Precautions

- Follow the manufacturer's guidelines in each step properly.
- Wear gloves while dealing with the reagents and plasticware.
- Avoid using an expired kit.
- Do not incubate the samples more than the recommended period of time; it may result in precipitation of salts.
- Load recommended volume of lysates; lower lysate volume may result in formation of air bubbles in tips resulting in problems of the liquid-handling system.
- Check for the presence of lysate and elution tube at corresponding positions.
- Check for the submerging of the entire sample in the lysis solution.
- Do not process the samples for extraction by PF Express™ and PF Express BTA™ kit simultaneously, as the liquid-handling system can process either of the protocol at a given point of time.
- Never remove the protocol card when the instrument is in "on" condition.
- Load the cartridges, lysates, tips, and elution tubes in their designated places.
- Do not touch the surface of the heat block as the temperature of the same could reach 95 °C.
- Carefully select the protocol you want to use in the liquid-handling system.
- Run the machine with proper power backup, as the process cannot be resumed in case of any power failure.
- Dispose the used cartridges and reagents carefully.
- Do not add acids or bases to the wastes containing lysis buffer as the acids/bases may react with guanidine thiocyanate in the lysis buffer and generate toxic gas.

8 Troubleshooting

Problem	Tentative cause	Possible solutions
Presence of precipitates in the liquid after lysis process	Extended time of incubation than the recommended conditions	Careful pipetting may dissolve the precipitate matter in the solution
	Lysates stored under refrigerated conditions	Vortex the lysate prior to centrifugation
	Ambient temperature very less	Avoid extended incubation of the samples during lysis process
		Avoid storing of the lysates under refrigerated conditions
No elution volume after run	No sample added to the sample tube	Rerun with proper samples and fresh reagent cartridges
	Sample volume lower than the recommended volume	Use recommended volume of sample as per the protocol used; if sufficient amount of lysate is not present make up the volume using Milli-Q
Colored DNA elute	Sample contains dyes	Perform qRT-PCR and check whether it affects C_T value of IPC or not
	DNA contaminated with heme	Minimize the amount samples, i.e., either blood or blood-stained clothing
Low yield of DNA	Protocol not followed properly or reagents not added during the lysis step	Review the protocol used and addition of appropriate reagents
	Incomplete lysis	Decrease the amount of input sample
		Add appropriate amount of proteinase-K
		Make sure to completely submerge the samples in lysis solution
		Incubation time may be increased

Chapter 12

Extraction of DNA by Using Anion-Exchange Resin Chelex®

Objective: To isolate DNA from liquid blood and forensic samples using Chelex® 100 resin.

1 Introduction

The primary step of forensic DNA analysis involves the extraction of high quality and quantity of DNA for their use in many downstream applications. Thus, it becomes highly essential to choose the most suitable technique for DNA extraction which is dependent on sample, technology, as well as analyst's preference. Additionally, other factors to be taken care of in selecting the suitable technique of DNA extraction include technical requirements, time efficiency, cost-effectiveness, as well as nature of biological samples to be processed.

In this regard, Singer-Sam in 1989 reported the use of Chelex 100 to increase the strength of PCR amplification from a low DNA tissue culture sample by simultaneous boiling. The initial work of Singer-Sam suggested the use of boiling suspensions of cells in 5% Chelex solution and use of the supernatant as template of PCR. The alkaline nature of the Chelex suspension (pH 10–11) and boiling of the samples at 100 °C generate the DNA fragments by disruption of the cell membranes. However, this method generates the denatured DNA samples which are not suitable for RFLP analysis.

The advantage of using Chelex is that it sequesters the divalent metals with the potential of introducing DNA damage. Additionally, Chelex beads can be removed from the lysate containing DNA without interfering the further downstream processing of the samples. As proteinase-K is not used in this DNA extraction practice, blood samples do not release heme from globin, thus generating less amount of free porphyrin compounds. This minimizes the inhibitor content in the extracted DNA. Less time requirement to

H. R. Dash et al., *Principles and Practices of DNA Analysis: A Laboratory Manual for Forensic DNA Typing*, Springer Protocols Handbooks, https://doi.org/10.1007/978-1-0716-0274-4_12,

complete the extraction procedure is another advantage of this procedure. Additionally, requirement of fewer manipulations in this isolation practice reduces the chance of inadvertent contamination of samples by extraneous DNAs.

2 Principle

Positively charged chemical substances such as anion-exchange resins bind to the oppositely charged substances such as DNA molecules. In this regard, Chelex constitutes the styrene divinylbenzene copolymers with iminodiacetate ions which chelates polyvalent metal ions, i.e., Mg^{2+}. Subsequently, the nucleases are inactivated and DNA is protected. The technique provides an added advantage in the purification of small fragments of DNA, mostly up to 1000 bp size. Hence, non-Chelex methods should be employed when high amount of DNA is expected from any sample.

Seligson and others in 1985 first described the use of anion-exchange resin materials for isolation of nucleic acid from varied sources such as whole blood. The discovery described the protocol of using a column containing positively charged resins of diethylaminoethyl cellulose groups on its surface for proper binding of negatively charged phosphate backbone of DNA. Additionally, the strength of binding of DNA as well as other impurities to the column can be optimized for salt concentration and pH. Other contaminants such as proteins and RNAs can be removed from the column by repeated washing using medium salt buffers.

3 Reagents Required and Their Role

3.1 Sodium Hypochlorite

Sodium hypochlorite (NaOCl) is the most widely used disinfectant and is known for its excellent cleaning action. However, the effectiveness of the solution is dependent on the concentration and pH of the solution. pH of the solution effects the formation of the hypochlorite ion ($^{-}$OCl) and proton (H^{+}) by dissociation of the weak acid hypochlorous acid (HOCl). In this case, HOCl plays an important role in germicidal action whereas the $^{-}$OCl determines the cleansing efficacy of the solution.

3.2 Chelex® 100 Resin

Chelex® 100 resin (Bio-Rad Laboratories, Hercules, CA, USA) consists of styrene divinylbenzene copolymers. These polymers also contain paired iminodiacetate ions. The role of this reagent during DNA extraction is the action of a chelating ion-exchange resin which binds polyvalent metal ions such as nucleases. This reagent has widespread application for DNA extraction with forensic implications.

4 Procedure

Preparation of Chelex® Mixture

1. With the help of 10% sodium hypochlorite, sterilize a dry spatula and a small magnetic bar.
2. Prepare 10% by weight of Chelex® 100 resin in sterilized HPLC-grade water.

   Note: Take a 50 ml sterile Falcon tube, add 5 g of Chelex® 100 resin into it, and make up the volume to 50 ml mark with sterilized HPLC-grade water.
3. Place the tube containing Chelex® 100 resin in water in a magnetic stirrer and mix well.
4. Transfer aliquots of 300 μl of 10% Chelex® 100 resin solutions to 1.5 ml sterile MCT and store in refrigerator.

DNA Extraction Step

5. Take out the required number of tubes containing 10% Chelex® 100 resin solutions from the refrigerator.
6. Label three tubes: one for blood, another for tissue sample, and third one for negative control.
7. Fill the hole of the heating block with water, and preset the heating block to 95 °C.
8. Sterilize forceps by flaming it over the alcohol burner several times. Using the sterilized forceps, remove a small portion of tissue (mostly of 0.2 mm × 0.2 mm size) from the sample and place it in the tube containing 10% Chelex® 100 resin solutions labeled for tissue sample.
9. Mix 100 μl of whole blood into another tube containing 10% Chelex® 100 resin solutions labeled for tissue sample.
10. Dip the sterilized forceps into the tube containing 10% Chelex® 100 resin solutions labeled for negative control.
11. Vortex the tubes containing the Chelex® solution and the samples.
12. Spin samples for 10–15 s at high speed to ensure that all the samples are dipped well into the 10% Chelex® 100 resin solutions.
13. Incubate the samples at 95 °C for 20 min in the heating block.
14. After incubation, vortex the samples for 10–15 s.
15. Centrifuge the tubes at 12,000 rpm for 5 min at 15 °C.
16. Transfer the supernatant into a fresh MCT and store at −20 °C till further use.

5 Observation

The quantity of the isolated DNA can be determined at the preliminary level by UV–visible spectrophotometer. For a 1 cm path length, the optical density at 260 nm (OD_{260}) equals 1.0 for the following samples:

- 50 µg/ml of double-stranded DNA
- 33 µg/ml of single-stranded DNA
- 20–30 µg/ml of oligonucleotide
- 40 µg/ml of RNA

6 Result Table

Sample	DNA content	$OD_{260/280}$	Inference
Control			
Sample 1 (*liquid blood*)			
Sample 2 (*tissue*)			

OD optical density

7 Precautions

- Wear gloves while dealing with the reagents and plasticware.
- During incubation, make sure that the lids of the tubes are closed properly to avoid pop off.
- Transfer the supernatant carefully; a little contamination of Chelex can inactivate Taq polymerase in the PCR setup.
- Always use an extra reaction for the negative control to check the level of contamination during the extraction process.
- It is recommended to use freshly prepared Chelex solution.
- Label the tubes appropriately.

8 Troubleshooting

Problem	Tentative cause	Possible solutions
No amplification of DNA after PCR	Inhibition of PCR reaction by leftover Chelex solution	Wait overnight before using the extracted DNA samples
		Dilute the extracted DNA sample to minimize the final content of leftover Chelex in the template
	Sufficient DNA not extracted	Repeat the extraction step
Mixed profile observed after genotyping	Contamination with exogenous DNA	Confirm contamination with negative control
		Use sterilized plasticware and glassware during the extraction process

Chapter 13

Isolation of DNA by Using Column-Based Extraction System

Objective: To isolate DNA from blood and forensic samples using column-based extraction system.

1 Introduction

Extraction of DNA is the fundamental method to be used in molecular biology and forensic DNA typing. DNA extraction from any biological material is either of analytical or preparative purposes. Earlier, many complicated, time-consuming, laborious techniques were used for isolation of DNA. However, with the advent of technology, many sophisticated, advanced, and specialized systems have been developed for extraction and purification of these biomolecules such as the solution- and column-based protocols. Gradually, the manual methods have been replaced by the semiautomatic and automated techniques which has not only reduced the time required for DNA extraction but also increased the quality of DNA obtained by reducing the chance of contamination.

The conventional DNA extraction steps involve the cell lysis to disrupt the cellular structure followed by inactivation of nucleic acid-degrading enzymes (i.e., DNase and RNase) and finally purifying the desired nucleic acid from the cell debris. This is called as liquid-liquid extraction of DNA. However, spin column-based nucleic acid purification can be achieved using a solid-phase extraction system for rapid purification of nucleic acid. This technique relies on the fact that nucleic acids bind to silica under specific conditions. Many column-based DNA extraction kits are available in the market. The selective use of the kits for forensic DNA typing application depends on the nature of the sample, type of sample, as well as analyst's preference.

H. R. Dash et al., *Principles and Practices of DNA Analysis: A Laboratory Manual for Forensic DNA Typing*, Springer Protocols Handbooks, https://doi.org/10.1007/978-1-0716-0274-4_13,

2 Principle

Column-based DNA extraction relies on the principle of solid-phase extraction system enabling the rapid purification of nucleic acids. The stages of this method involve lysis, binding, washing, and elution for effective purification of DNA. During lysis, the cells are broken open from cellular as well as nuclear membrane to release the nucleic acid into the solution. In the binding step, a binding solution is applied on the spin column to provide it a specific condition for categorical binding of nucleic acids. Upon addition of the lysed product to the spin column followed by centrifugation, at optimal pH and salt concentration, the nucleic acid binds to the silica gel membrane and other biomolecules present in the lysis solution pass through. This step is followed by the addition of wash buffer which removes the remaining cell debris leaving only DNA to bind to the silica membrane. Finally, an elution buffer is added to the column which removes the nucleic acid from the column for its subsequent collection at the bottom of the tube.

The major composition of these silica membranes is diethylaminoethanol (DEAE). Under suitable pH, when H^+ ions are added to the column, it becomes positively charged. As it is a well-known fact that DNA molecule is negatively charged, it can categorically bind to the positively charged column and other cell debris pass through it (Fig. 13.1). This enables the purification of DNA optimally with highest quantity and quality.

There are many commercially available kits in the market for column-based extraction of DNA from forensic samples. These kits may be used manually and employed along with the semiautomatic instruments to minimize repeated pipetting. In this section we have included automated extraction of DNA from liquid blood sample using QIAamp® DNA Investigator® kit and QIAcube liquid-handling system (QIAGEN®). The other available kits with forensic DNA isolation implications employing column-based principle have been listed in Appendix B of the current book.

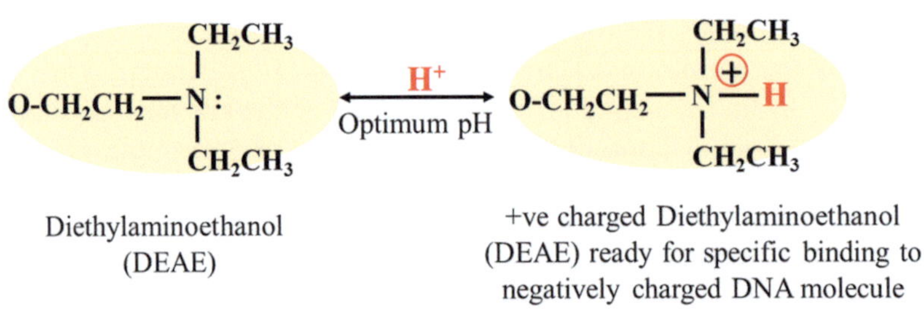

Fig. 13.1 Principle of binding of silica columns to negatively charged DNA for effective purification

3 Reagents and Plasticware Required and Their Role

3.1 QIAamp® DNA Investigator® Kit

The constituents of the kit include but not limited to the following items:

Buffer AL: It is used for isolation of DNA using the aforementioned kit, mostly for lysis of the cells. Buffer AL may precipitate upon storage, which can be removed by its incubation at 55 °C prior to the start of DNA isolation. However, buffer AL can remain stable at room temperature (15–25 °C) till 1 year.

Buffer AW1: This represents the first wash buffer. Buffer AW1 generally comes in the form of concentrated solution which needs to be diluted prior to use. 19 ml of buffer should be diluted with 25 ml of absolute ethanol. This reconstituted buffer can be stored at room temperature and used up to 1 year. However, this buffer should be mixed by shaking before use.

Buffer AW2: This represents the second wash buffer. Like AW1, 13 ml of AW2 should be reconstituted with 30 ml of absolute ethanol. The reconstituted AW2 will remain stable at room temperature for 1 year. Buffer should be mixed by shaking prior to its use.

Carrier RNA: Carrier RNA is supplied with the kit. It enhances the binding of DNA to the membrane of QIAamp MinElute® columns. It is useful in low-DNA samples when less amount of target molecules are available to bind the column. However, carrier RNA also gets extracted along with the DNA in the final elution step. Thus, carrier RNA should be added wisely and for samples with higher amount of DNA carrier RNA should not be used.

Proteinase-K: Peptide bond present near the carboxyl group of amino acids with blocked alpha amino groups is the cleavage site for the serine protease, i.e., proteinase-K. It helps in degrading many protein impurities to yield a good quality of DNA. It also plays an important role in the inactivation of nucleases, hence preventing the isolated DNA from damage. The activity of proteinase-K is greatly increased by its simultaneous action with SDS and high temperature. Proteinase-K works optimally at 50–65 °C as higher temperature unfolds certain proteins to ease the function of the enzyme. However, proteinase-K is being supplied by the kit manufacturer.

QIAamp MinElute® Columns: These columns work on the fast spin column procedure, where the traditional phenol-chloroform extraction steps are not required. In this case, the nucleic acid categorically binds to the membrane of the columns and allows the other contaminants to pass through. Washing of the column membranes allows the efficient removal of many PCR inhibitors such as divalent cations and proteins.

Buffer ATE: This buffer helps in the efficient removal of DNA from the bound columns and maintains proper storage condition of the eluted DNA. In an alternative, molecular biology-grade water can also be used for elution of DNA from the columns. The elution volume is chosen significantly as per the nature and condition of sample as well as the requirements of the downstream processing of the samples.

4 Procedure

1. Transfer a minimum of 1 μl and a maximum of 100 μl of liquid whole blood sample to a 1.5 ml microcentrifuge tube.
2. Make up the total volume to 100 μl by adding buffer ATL.
3. Add 100 μl of buffer AL and 10 μl of proteinase-K, vortex for 15 s to prepare a homogenized solution, and incubate the samples at 56 °C for 10 min.
4. Add 50 μl of absolute ethanol to the lysate and mix properly by vortexing for 15 s.
5. Incubate the tubes at room temperature (15–25 °C) for 2–5 min.
6. Prepare the QIAamp MinElute® column by placing it in a 2 ml collection tube and without wetting its rim transfer the entire lysate to the column.
7. Centrifuge the column containing the lysate at 6000 × *g* for 1 min at room temperature.
8. Discard the collection tube containing the flow through and place the QIAamp MinElute® Column in a fresh 2 ml collection tube.
9. Add 500 μl buffer AW1 to the column and centrifuge at 6000 × *g* for 1 min at room temperature.
10. Discard the collection tube containing the flow through and place the QIAamp MinElute® Column in a fresh 2 ml collection tube.
11. Add 700 μl of buffer AW2 to the column and centrifuge at 6000 × *g* for 1 min at room temperature.
12. Discard the collection tube containing the flow through and place the QIAamp MinElute® Column in a fresh 2 ml collection tube.
13. Add 700 μl of absolute ethanol to the column and centrifuge at 6000 × *g* for 1 min at room temperature.

14. Discard the collection tube containing the flow through and place the QIAamp MinElute® Column in a fresh 2 ml collection tube.
15. Centrifuge at full speed for 3 min at room temperature to remove excess ethanol remaining in the column.
16. Place the QIAamp MinElute® Column in a fresh 2 ml collection tube and incubate at room temperature for 10 min by opening the lid of the column.
17. Add 20–50 μl of buffer ATE to the center of the membrane and incubate for 1 min at room temperature.
18. Centrifuge at full speed for 1 min. Transfer the elute from the collection tube to a fresh 1.5 ml microcentrifuge tube and store at −20 °C.

5 Observation

The quantity of the isolated DNA can be determined at the preliminary level by UV–visible spectrophotometer. For a 1 cm path length, the optical density at 260 nm (OD_{260}) equals 1.0 for the following samples:

- 50 μg/ml of double-stranded DNA
- 33 μg/ml of single-stranded DNA
- 20–30 μg/ml of oligonucleotide
- 40 μg/ml of RNA

6 Result Table

Sample	DNA content	$OD_{260/280}$	Inference
Control			
Sample 1 (*liquid blood*)			
Sample 2 (*coagulated blood*)			

OD optical density

7 Precautions

- Follow the manufacturer's guidelines in each step carefully.
- Wear gloves while dealing with the reagents and plasticware.
- Avoid using an expired kit.

- Do not incubate the samples more than the recommended period of time; it may result in precipitation of salts.
- Load the lysate to the QIAamp MinElute column without wetting its rim.
- Change pipette tips in between two steps.
- Use aerosol-resistant tips to avoid contamination.
- Do not touch the QIAamp MinElute column membrane with tip.

8 Troubleshooting

Problem	Tentative cause	Possible solutions
Less/no DNA elute	Carrier RNA not added	Dissolve carrier RNA to buffer AL for low DNA sample and repeat the purification step
	Improper storage of samples	Avoid repeated freezing and thawing of samples; use recommended storage condition of the samples
	Improper lysis of samples	Monitor the storage of proteinase-K; use a new batch of proteinase-K
Poor quality of DNA	pH of water used for elution was too low	DNA will not dissolve properly in acidic solutions; ensure correct pH of water
Clogged QIAamp MinElute column	Incomplete lysis of sample	Increase lysis time or proteinase-K concentration

Chapter 14

Reliable Use of Whatman™ FTA™ Cards for One-Step Collection and Isolation of DNA

Objective: One-step collection and isolation of DNA from varied sources using Whatman™ FTA™ cards and the technology.

FTA™ (Flinders Technology Associates) card (GE Whatman, Maidstone, Kent, United Kingdom) is a specialized filter paper which is impregnated with reagents promoting cell lysis and protein denaturation followed by release of nucleic acids that are entrapped in the card matrix and stable at room temperature for a long period of time. Though forensic DNA technology has geared up over the years in terms of automation and many up-gradation and standardized protocols are in practice, DNA extraction and sample collection steps still remain the analyst's preference. In this regard, FTA™ card may prove to be useful for a wide range of forensic samples.

These cards enable the optimal collection, transportation, purification, as well as long-term storage of DNA at room temperature by the use of a single card, which can be used for both forensic and database applications. The fundamental of FTA™ cards is that it lyses the eukaryotic cells that are in contact with the cards. After lysis the proprietary chemical content of the card protects the DNA from the environmental and enzymatic damages. Additionally, DNA on FTA™ cards need not be stored at freezing temperature as DNA on FTA™ cards has been evidenced to remain stable at room temperature. Many countries have provisions for long-term storage and retesting of DNA samples after several years such as post-conviction testing. In such cases storage of DNA on FTA™ cards is highly useful. Additionally, the chemicals present on these cards also protect the DNA from harmful UV lights. US Postal Service Mail has marked FTA™ cards to be safe, and thus can be transported without any hazardous labeling.

H. R. Dash et al., *Principles and Practices of DNA Analysis: A Laboratory Manual for Forensic DNA Typing*, Springer Protocols Handbooks, https://doi.org/10.1007/978-1-0716-0274-4_14, © Springer Science+Business Media, LLC, part of Springer Nature 2020

1 Sample Types and Collection Procedures

The cards have been designed to be used for collection, transport, and storage of a wide range of samples, viz. buccal cells, saliva, blood, and other soft tissues. The ideal procedure for collection of samples has been described below.

1.1 Collection of Liquid Blood

1. Wear gloves, and label the cards with appropriate sample identification.
2. Drop <125 μl of freshly collected liquid blood (per 1 in. circle) in a concentric circular motion on the card.
3. Dry the card for 3–5 min.
4. Place the card in a sealed envelope and send it for investigation at room temperature.

1.2 Collection of Buccal Cells

1. Roll the Catch-All sample collection swabs (Epicentre Biotechnologies) tightly inside both sides of the cheek for 15 s.
2. Rub the swabs in a concentric circular motion on the card.
3. Dry the card for 3–5 min.
4. Place the card in a sealed envelope and send it for investigation at room temperature.

1.3 Collection of Saliva Samples

1. Ask the sample donor to wash his/her mouth with water and wait for 30 s.
2. Tell him/her to spit in the blue container (DNA Genotek).
3. Dip the Catch-All sample collection swabs (Epicentre Biotechnologies) in the container containing saliva and rub it in a concentric circular motion on the card.
4. Dry the card for 3–5 min.
5. Place the card in a sealed envelope and send it for investigation at room temperature.

Note: Store the unused cards in a cool, dry, and clean environment. Avoid puddling of the samples. Place the sample within the marked circular area. Do not rub or smear the sample on the card.

2 DNA Extraction and Usages (Applications)

Though many genotyping kits (PowerPlex® ESX, AmpFlSTR Identifiler Direct PCR Amplification Kit, GlobalFiler® Express, PowerPlex® Fusion 6C, Investigator® 24plex GO!, and AmpFISTR® Yfiler® Plus to mention a few) have been developed for direct amplification of DNA sample on FTA™ cards, DNA can also be extracted from FTA™ cards using any of the following methodologies.

2.1 Organic Extraction

1. Remove a disk of around 2 mm size from the center of a FTA™ card containing dried blood stains using a Harris Uni-Core disposable punch.
2. Place the disk in a 1.5 ml DNase/RNase-free microcentrifuge tube.
3. Add 500 μl of extraction buffer (10 mM Tris–HCl, pH 8.0; 10 mM EDTA, disodium salt, pH 8.0; 100 mM sodium chloride; and 2% v/v SDS) and 20 μl of proteinase-K (20 mg/ml).
4. Mix by vortexing and incubate at 56 °C for 2 h in a water bath with gentle shaking.
5. Add equal volume of phenol (pH 8.0), mix properly, and centrifuge at maximum speed for 10 min at 4 °C.
6. Transfer the aqueous upper layer to a fresh 1.5 ml microcentrifuge tube and add equal volume of chloroform to it.
7. Mix properly and centrifuge at maximum speed for 10 min at 4 °C.
8. Transfer the upper aqueous phase to a fresh 1.5 ml microcentrifuge tube and add 50 μl 3 M sodium acetate (pH 5.2) to it, followed by the addition of 800 μl of 100% ethanol, mixing by vortexing, and precipitation at −20 °C for 1.5 h.
9. Centrifuge for 30 min at maximum speed at 4 °C and discard the supernatant.
10. Add 1 ml of 70% ethanol and centrifuge at maximum speed for 20 min at 4 °C.
11. Dry the pellet at room temperature followed by addition of 50 μl of TE buffer for dissolving of pellet.
12. Store the extracted DNA at −20 °C till further use.

2.2 Using Chelex 100 Resin

1. Remove a disk of around 2 mm size from the center of a FTA™ card containing dried blood stains using a Harris Uni-Core disposable punch.
2. Place the disk in a 1.5 ml DNase/RNase-free microcentrifuge tube.
3. Crush the disk using a 20-gauge needle.
4. Wash the crushed disk twice by adding 1 ml of sterile distilled water followed by incubation at room temperature for 10 min and removal of water.
5. Centrifuge the tube at maximum speed for 5 min and discard the supernatant.
6. Add 200 μl of Chelex 100 to the tube and incubate at 56 °C for 20 min.
7. Mix by vortexing for 15 s and incubate at 100 °C for 8 min followed by mixing by vortexing for 15 s.

8. Centrifuge the tube at maximum speed for 5 min, transfer the supernatant containing extracted DNA to a fresh tube, and store at −20 °C till further use.

2.3 QIAamp™ DNA Investigator Kit

1. Remove a disk of around 2 mm size from the center of a FTA™ card containing dried blood stains using a Harris Uni-Core disposable punch.
2. Place the disk in a 1.5 ml DNase/RNase-free microcentrifuge tube.
3. Add 280 μl of buffer ATL to the tube, add 20 μl of proteinase-K to it, and mix by vortexing for 30 s.
4. Incubate at 56 °C for 2 h in water bath with gentle shaking.
5. Centrifuge the tube at maximum speed for 30 s and add 300 μl of buffer AL to it.
6. Mix by vortexing and incubate at 70 °C for 10 min in a water bath with gentle shaking.
7. Centrifuge the tube at maximum speed for 30 s and transfer the supernatant to the minElute column.
8. Centrifuge the column at 6000 × g for 1 min at room temperature and discard the flow through.
9. Add 700 μl of buffer AW2 to the column, centrifuge at 6000 × g for 1 min at room temperature, and discard the flow through.
10. Add 700 μl of 100% ethanol to the column, centrifuge at 6000 × g for 1 min at room temperature, and discard the flow through.
11. Centrifuge the column at maximum speed for 3 min and incubate the column at room temperature for 10 min by opening the lid of the column.
12. Add 50 μl of sterile distilled water/buffer ATE and incubate at room temperature for 5 min.
13. Centrifuge at maximum speed for 1 min, and store the flow through containing DNA at a fresh centrifuge tube at −20 °C till further use.

2.4 illustra™ Tissue and Cells genomicPrep Mini Spin Kit

1. Remove a disk of around 2 mm size from the center of a FTA™ card containing dried blood stains using a Harris Uni-Core disposable punch.
2. Place the disk in a 1.5 ml DNase/RNase-free microcentrifuge tube.
3. Add 1 ml of sterile phosphate-buffered saline (PBS) to the tube and centrifuge at 16,000 × g for 1 min at room temperature.

4. Crush the disk using a 20-gauge needle; centrifuge at 2000 × *g* for 10 s at room temperature.
5. Add 50 μl of buffer 1 to the tube, add 10 μl of proteinase-K, and vortex for 15 s.
6. Incubate at 56 °C for 1 h.
7. Centrifuge at 2000 × *g* for 10 s; add 5 μl of RNase A (20 mg/ml) followed by incubation at room temperature for 15 min.
8. Add 500 μl of buffer 4 and mix by vortexing for 15 s, followed by incubation at room temperature for 10 min.
9. Transfer the whole content (lysate) to a mini column, centrifuge at 11,000 × *g* for 1 min, and discard the flow through.
10. Add 500 μl of buffer 4 to the column, centrifuge at 11,000 × *g* for 1 min, and discard the flow through.
11. Place the column on a fresh collection tube, add 500 μl of buffer 6, and centrifuge at 11,000 × *g* for 3 min.
12. Transfer the column to a fresh 1.5 ml microcentrifuge tube, add 50 μl pre-warmed elution buffer to the column, and incubate at room temperature for 1 min.
13. Centrifuge at 11,000 × *g* for 1 min and store the elute containing DNA at −20 °C till further use.

2.5 DNA IQ™ Kit

1. Remove a disk of around 2 mm size from the center of a FTA™ card containing dried blood stains using a Harris Uni-Core disposable punch.
2. Immerse the punch with 100 μl of lysis buffer and incubate at 70 °C for 30 min.
3. Transfer the lysate to a column placed in a 1.5 ml microcentrifuge tube and centrifuge for 2 min at maximum speed at room temperature.
4. Collect the flow through and add 7 μl of premixed resin to it.
5. Mix by vortexing for 3 s followed by incubation at room temperature for 5 min with repeated vortexing.
6. Place the tube on a magnetic stand, and discard the supernatant without disturbing the pellet.
7. Add 100 μl of lysis buffer, vortex the tube for 2 s, place the tube on magnetic stand, and discard the lysis buffer.
8. Add 100 μl of 1× wash buffer, vortex the tube for 2 s, place the tube on magnetic stand, and discard the wash buffer.
9. Repeat step 8 two more times.
10. Place the tube on a magnetic stand with lid open and air-dry for 10 min.

11. Add 100 μl of elution buffer, vortex for 2 min, and incubate at 65 °C for 5 min.
12. Vortex for 2 s, place the tube on magnetic stand, and transfer the elute to a fresh tube.
13. Store the tube containing DNA at −20 °C till further use.

FTA or any other commercially available cards can have a wider application in forensic use which includes but not limited to the following: DNA extraction, DNA quantitation using real-time PCR, STR analysis using conventional genotyping kits, analysis of mitochondrial DNA, as well as SNP analysis.

3 Available FTA™ Cards and Other Related Products

Many types of FTA™ cards are available in the market. Additionally, similar cards from different manufacturers are also available commercially which has been listed in Table 14.1.

Table 14.1
Commercially available cards for direct collection and extraction of DNA from biological samples and their nature

Products	Feature	Manufacturer and address
FTA Classic Card	125 μl/sample area	GE Healthcare, UK
Indicating FTA Classic Card	125 μl/sample area	GE Healthcare, UK
FTA Mini Card	125 μl/sample area	GE Healthcare, UK
Indicating FTA Mini Card	125 μl/sample area	GE Healthcare, UK
FTA Micro Card	125 μl/sample area	GE Healthcare, UK
Indicating FTA Micro Card	125 μl/sample area	GE Healthcare, UK
FTA Gene Card	75 μl/sample area	GE Healthcare, UK
PlantSaver™ Card	25 μl/sample area	GE Healthcare, UK
CloneSaver™ Card	5 μl/sample area	GE Healthcare, UK
EasiCollect	125 μl/sample area	GE Healthcare, UK
FTA Elute Micro Card	30 μl/sample area	GE Healthcare, UK
FTA Elute Micro Card	30 μl/sample area	GE Healthcare, UK
Ahlstrom-Munksjö GenSaver™	125 μl/sample area	AHLSTROM-MUNKSJÖ, North America
Ahlstrom-Munksjö GenSaver™ Color	70 μl/sample area	AHLSTROM-MUNKSJÖ, North America
Ahlstrom-Munksjö GenSaver™ 2.0	5/125 μl/sample area	AHLSTROM-MUNKSJÖ, North America

(continued)

Table 14.1 (continued)

Products	Feature	Manufacturer and address
Ahlstrom-Munksjö GenSaver™ Color 2.0	5/70 μl/sample area	AHLSTROM-MUNKSJÖ, North America
NUCLEIC-CARD™	N/A	COPAN ITALIA S.P.A., Italy
GENO•CARD	N/A	Hain Lifescience GmbH, Germany
NucleoCard®	<200 μl	MACHEREY-NAGEL, Germany
GenSaver™ DNA cards	N/A	Gen*Tegra*®, Pleasanton, CA
NUCLEIC-CARD™ matrix	N/A	Thermo Fisher Scientific
InstaDNA™ Card	10–200 μl	HiMedia, Mumbai, India

Part IV

Qualitative and Quantitative Assessment of Extracted DNA

Chapter 15

Quantification of DNA by Using Agarose Gel Electrophoresis Technique

Objective: To quantify the isolated DNA from blood and forensic samples using agarose gel electrophoresis technique.

1 Introduction

Downstream processing of samples requires a defined concentration of DNA, and hence proper quantification of DNA is the must prerequisite during DNA analysis. DNA content of forensic samples varies greatly with the nature of sample, storage condition, as well as their stages of degradation. Hence, proper quantification of the same is highly essential prior to any further analysis. Though many techniques are available for suitable quantification of DNA, agarose gel electrophoresis is considered to be the most suitable physical technique for the same. In this technique, DNA migrates along an agarose matrix under the influence of electric current. Many factors control the migration of DNA on agarose gel which include size of DNA, concentration of agarose matrix, voltage applied, presence or absence of ethidium bromide, type and nature of agarose, as well as strength of buffer used. It is a typical sieving process where the small-sized DNA fragments migrate faster and the large fragments migrate slower which is dependent on the agarose concentration.

The sensitivity of detection of DNA fragments during STR analysis is dependent on the sensitivity of the DNA quantification methodology used. In this regard, electrophoresis through agarose gel provides an added advantage as it can separate, identify, and purify nucleic acids. The matrix of the agarose gel has adjustable and regular pore size. Chemically inert nature of agarose gel also provides an added advantage.

H. R. Dash et al., *Principles and Practices of DNA Analysis: A Laboratory Manual for Forensic DNA Typing*, Springer Protocols Handbooks, https://doi.org/10.1007/978-1-0716-0274-4_15,

2 Principle

Agarose is a natural linear polymer which is extracted from the seaweed. It can form a matrix by hydrogen bonding when cooled followed by heating in a buffer. These gels are simple in nature and can be prepared rapidly. Moderate- and large-sized nucleic acids can be separated using this medium with optimal resolution depending on the concentration of the gel. The concentration of the agarose gel to be prepared depends on the fragment size of DNA. Mostly the gel is prepared with a concentration of 0.2–3% with the fundamentals that lower the agarose gel concentration and increase the migration of DNA fragments. For a proper resolution of banding patterns of the DNA fragments, a standard concentration of agarose gel should be used as per Table 15.1.

The basic principle of agarose gel electrophoresis includes the characteristic migration of a charged molecule towards either positive or negative electrode under the influence of electric field (Fig. 15.1). Nucleic acids possess a constant negative charge due to the net charge imparted by their phosphate backbone constituents and hence migrate towards the anode. A potential gradient "*E*" is generated under the influence of the voltage applied which can be determined by the equation, $E = V/d$ (where "*V*" is the amount of voltage applied and "*d*" is the distance between the electrodes in centimeter). Under the influence of the electric gradient "*E*," a force is generated on the charged molecule which can be

Table 15.1
Concentrations of agarose gel and their respective power of resolutions for linear DNA molecules

Agarose concentration (in %)	Size of DNA fragments (in kb)
0.2	5–40
0.3	5–60
0.4	5–30
0.6	1–20
0.7	0.8–10
0.8	1–7
0.9	0.5–7
1	0.5–5
1.2	0.4–6
1.5	0.2–3
2	0.1–2
3	0.1–1

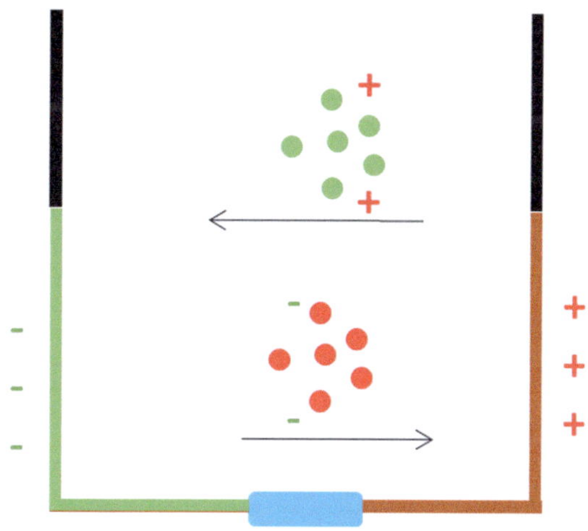

Fig. 15.1 Migration pattern of charged molecules under the influence of electric field

expressed by $F = E{\cdot}q$ (where "q" is the charge of the molecule in coulombs). This force "F" is responsible for dragging of the DNA molecule to the corresponding electrode as per the charge of the molecule. Additionally, a negative force also exists which helps in slowing down of the migration pattern of the molecule called as frictional force and is dependent on (a) hydrodynamic size of the molecule, (b) shape of the molecule, (c) pore size of agarose gel, and (d) viscosity of the buffer used.

During analysis of DNA fragments on agarose gel, a fluorescent intercalating dye is added to the gel matrix, i.e., ethidium bromide or gold view. These dyes intercalate to the DNA molecules as they migrate in the gel and flourish under UV light. The principle of quantification of DNA molecules includes the fact that the higher the amount of DNA molecules, the more the intensity of bands visualized under UV light. Thus, when a standard concentration of DNA molecule is run on agarose gel side by side with the unknown DNA fragment, by comparing the intensity of the bands, the relative as well as absolute quantification of the DNA molecule can be carried out.

3 Reagents Required and Their Role

3.1 Agarose

Agarose consists of the repeated subunits of agarobiose, i.e., L- and D-galactose subunits. It is mostly extracted from the seaweeds of *Gelidium* and *Gracilaria* genera. During gelation the polymers of agarose associate with each other non-covalently to form a macroporous matrix with specific pore size for the specific separation of nucleic acids. It is nontoxic, devoid of free radical involvement during polymerization, and thermoreversible in nature which are the added advantages of agarose for separation of nucleic acids during electrophoresis.

3.2 Tris-Acetate Buffer

The mobility of DNA is dependent on the composition and ionic strength of the electrophoresis buffer. In this regard, two most common buffers are employed which include Tris-borate-EDTA and Tris-acetate-EDTA. The electrophoresis buffer maintains the pH of the reaction to neutral, and thus net charge of DNA solely becomes responsible for its migration under the influence of electric field. Additionally, buffer provides a constant liquid medium to prevent hydrolysis of DNA molecules besides preventing DNA from the attack of DNase. It is always recommended to use freshly prepared 1× TAE buffer during electrophoresis. However, a stock solution of 10× TAE can be prepared using the following composition: Tris—400 mM, acetic acid—200 mM, EDTA—10 mM, and distilled water—up to 250 ml.

3.3 Ethidium Bromide

Ethidium bromide (EtBr) is a ring-structured hydrophobic compound which resembles the nucleotide bases of DNA. It is an intercalating agent and intercalates between the stacked nucleotide bases of DNA by forming close van der Walls interaction with the base pairs. Ethidium is fluorescent in nature and it flourishes in visible range when exposed to UV light. Thus during electrophoresis, the DNA fragments that contain intercalated EtBr molecules also flourish without which the DNA fragments would not have been visualized. EtBr is required in a trace amount (0.5 μg/ml) during electrophoresis. For this, add 5 μl of stock EtBr solution (concentration 10 mg/ml) after the molten agarose gel has cooled down to 60–70 °C.

3.4 Gel-Loading Buffer

Gel-loading buffer comprises dyes and other compounds that help in assessing the speed of the DNA sample during electrophoresis as well as ascertaining that the sample becomes denser than the buffer. The dye varies with the manufacturer; however, xylene cyanol, cresol red, bromophenol blue, orange G, and tartrazine are the common constituents. Additionally, to increase the density of DNA, another constituent is generally present in the gel-loading buffer which includes Ficoll, sucrose, or glycerol. Mostly, the commercially available gel-loading buffer comes as 6× concentration and its final working concentration should be 1×. The buffer should be stored at 4 °C to avoid the growth of molds utilizing sucrose.

3.5 DNA Size Marker

Most of the manufacturers provide ready-to-load DNA size markers. The concentration of size marker varies with the manufacturer and during loading manufacturer's recommended guidelines should be followed. A DNA ladder of 1 kb size generally gives fragments of various sizes, i.e., 0.5, 1.0, 1.5, 2.0, 3.0, 4.0, 5.0,

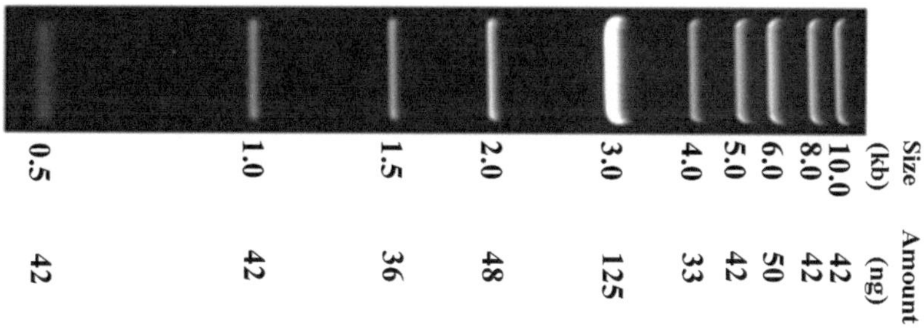

Fig. 15.2 Different fragments of DNA in a typical DNA size marker along with their corresponding concentration of DNA

6.0, 8.0, and 10.0 kb. The amount of DNA (in ng) present on each DNA fragment is given along with the datasheet of the manufacturer which is typically as follows (Fig. 15.2):

3.6 DNA Sample

Extracted DNA from reference as well as forensic samples can be used for estimation of DNA quantity. If you expect a higher quantity of DNA, dilute the extract in TE buffer prior to running the samples on agarose gel. Various known concentrations of control DNA can be prepared in TE buffer for assessment of measurement of DNA concentration by check gel technique.

3.7 Gel Documentation System and Software for Analysis

Many gel documentation systems are available in the market and they have their own analysis software for banding pattern analysis. For example Bio-Rad™ has the gel documentation system with the analysis software Quantity One®.

4 Procedure

1. Mix 30 mg of agarose in 30 ml of 1× TAE in a conical flask to prepare 1% mixture.
2. Use a plastic to cover the flask. Pierce a small hole in the plastic for ventilation.
3. Boil the solution using a microwave, and swirl the container gently to resuspend the settled agarose.
4. Cool the agarose till you can touch the flask and add ethidium bromide at the final concentration of 0.5 μg/ml (in this case add 1.5 ml from 10 mg/ml stock solution).
5. Pour the gel mixture in the gel apparatus using the gel-casting tray and comb.
6. Carefully remove the comb after solidification.
7. Prepare a written record of the samples loaded on each well. Mix 10 μl of DNA sample and 2 μl of gel-loading dye and load the final 12 μl to each well.

8. Fill the electrophoresis tank with 1× running buffer, cover the chamber with lid, and connect the electrodes with power supply.
9. Turn on the power supply; monitor the gel carefully; once the dye moves through the gel disconnect the electricity.
10. View the DNA sample in the gel documentation system.
11. Using the analysis software compare the intensity of the unknown DNA fragments with the intensity of known DNA fragments of the ladder.
12. The software will automatically calculate the DNA concentration of the unknown sample.

5 Observation

Manually observe the bands on the gel slab under UV light. Presence of various bands in the lane with DNA ladder confirms no methodological error. Prepare standard curve using the fluorescence intensity vs. DNA amount in each band and calculate the quantity of the isolated DNA using the same standard curve.

6 Result Table

Sample	Fluorescence intensity	Amount of DNA
Control		
Sample 1 (*liquid blood*)		
Sample 2 (*bone powder*)		

7 Precautions

- Wear actual-fit heat-resistant gloves for preparation and pouring of molten agarose.
- During manual checking of bands under UV light always use UV-proof full-face shield.
- Clean the glassware thoroughly before preparing the agarose gel.
- Do not overboil the agarose in the hot plate or microwave.
- Handle EtBr with extra care as it is highly carcinogenic in nature.
- Always use freshly prepared buffer for preparation of agarose and running buffer.
- Dispose the used gels carefully.

8 Troubleshooting

Problem	Tentative cause	Possible solutions
Faint of no band on the gel	DNA concentration very low	Increase the loading amount of DNA in the gel; if problem persists re-isolate DNA
	Highly degraded DNA	Re-isolate DNA with extra precaution for nuclease contamination
	DNA migrated out of the slab gel	Decrease the electrophoresis time; limit electrophoresis voltage to 20 V/cm; temperature during electrophoresis should be maintained at <30 °C
	UV light source not proper	Use UV light source of wavelength 254 nm
No band, only smear observed	Highly degraded DNA	Re-isolate DNA with extra precaution for nuclease contamination
	Too much DNA	Dilute the sample with TE buffer prior to loading on agarose gel
	Electrophoresis conditions not proper	Maintain electrophoresis voltage to 20 V/cm and temperature during electrophoresis <30 °C

Chapter 16

Quantification of DNA by Using UV–Visible Spectrophotometer

Objective: To quantify DNA isolated from blood and forensic samples using UV–visible spectrophotometer technique.

1 Introduction

Quantification of DNA from the reference as well as the forensic (questioned) samples is of utmost importance during DNA fingerprinting analysis. Though many advanced techniques are available nowadays, UV–visible spectrophotometer is the most fundamental technique of DNA quantification. This is mostly carried out by measuring the solution containing DNA at 260 nm which is the absorption maximum for DNA. This technique is very simple and time saving in nature. However, high interference in result is observed due to the presence of degraded nucleotides, denatured nucleic acids, and colored compounds which are common in DNA extracted from forensic samples.

In spite of all pitfalls, UV–visible spectrophotometer-based technique is the most widely used technique for DNA quantification due to its simplicity in experimentation and analysis of obtained results. Additionally, this technique allows to assess the quality of extracted DNA by measuring the protein and/or RNA contamination in the sample. Compounds used during nucleic acid preparation can also show absorbance at 260 nm, leading to inaccurate measurement of DNA quantity. However, this interference from these non-DNA samples can be avoided by measuring and calculating the ratio of absorbance at A_{260}/A_{280}.

H. R. Dash et al., *Principles and Practices of DNA Analysis: A Laboratory Manual for Forensic DNA Typing*, Springer Protocols Handbooks, https://doi.org/10.1007/978-1-0716-0274-4_16,

2 Principle

The most common method of measurement of DNA concentration in an aqueous solution is to measure its absorbance at 260 nm, as the absorption maximum of DNA falls at this UV wavelength. In this regard, the absorbance is measured using a spectrophotometer and the DNA concentration is measured using the Beer-Lambert law. This law deciphers the relationship between absorbance and concentration of DNA in a linear fashion, which can be described using the following equation:

$$A = \mathrm{OD} = \varepsilon \times l \times c$$

where A: absorbance, OD: optical density, ε: molar extinction coefficient, l: path length of cuvette, and c: concentration of the solution.

The amount of extract to be used for measurement of DNA concentration lies with the capacity of cuvette used. When the capacity of cuvette is <0.2 ml, then 5 μl of extract mixed with 195 μl of water should be used to measure the absorbance. The absorbance should be measured in triplicates to get the average result. It is always recommended to discard the value of absorbance at 260 nm either lower than 0.02 or between 1 and 1.5 due to chance of error at a higher rate. Finally, the concentration of the target molecule can be calculated using the following formula:

$$\text{Concentration (pmol/μl)}: A_{260}/0.027 \text{ [for single-stranded DNA]}$$
$$\text{Concentration (pmol/μl)}: A_{260}/0.020 \text{ [for double-stranded DNA]}$$
$$\text{Concentration (pmol/μl)}: A_{260}/0.025 \text{ [for single-stranded RNA]}$$

Note: For a cuvette with 1 cm path length, an absorption at 260 nm $A = 1$ corresponds to approximately 50 μg/ml of double-stranded DNA, 37 μg/ml of single-stranded DNA, 40 μg/ml of RNA, or 30 μg/ml of oligonucleotides. Single-stranded DNA absorbs more UV than double-stranded DNA evidenced by the increase in OD value. This change in OD value between single- and double-stranded DNA is called as hyperchromic shift.

In addition to measuring the quantity of DNA, it is equally important to measure the quality of DNA prior to further downstream processing of samples. Proteins, phenols, salts, and other compounds routinely used during DNA extraction process can also interfere with the proper quantification of DNA by absorbance technique. Purity of nucleic acid samples can be determined by measuring the ratio of absorbance at 260 nm (A_{max} of DNA) and 280 nm (A_{max} of protein). Optionally, background correction can be done by measuring absorbance at 320 nm at which neither DNA nor protein absorbs any light. The approximate value of purity of the desired compounds can be determined by using the following ratios:

Type	A_{260}/A_{280} (approximate)
Pure DNA	1.8
Pure RNA	2.0
Pure protein	0.57

3 Reagents/Materials/Instruments Required and Their Role

3.1 Spectrophotometer

Spectrophotometer is used to determine the concentration of a solute by transmitting light through the solution. The basic principle of the instrument lies with the fact that when a known wavelength of light is passed through a cuvette containing sample, the amount of light transmitted is measured to calculate the absorbance by using a photocell. The simplest form of a spectrophotometer involves a light source, a prism to filter a specific wavelength of light, a sample holder, and a photocell (Fig. 16.1).

3.2 Cuvette

The choice of cuvette to be used depends on the amount of nucleic acid solution to be measured. Generally, the size and capacity of the cuvette vary from 1 ml to as low as 5–70 μl. Thus, the cuvette should be selected based on the sample concentration range, dilution factor, and available sample volume.

3.3 TE Buffer

TE buffer is generally prepared by mixing 50 mM Tris and 50 mM EDTA at pH 8.0. Tris, the major constituent of TE buffer, acts as a common pH buffer. Additionally, EDTA chelates cations like Mg^{2+}. Hence, TE buffer is helpful to solubilize DNA by protecting it from degradation.

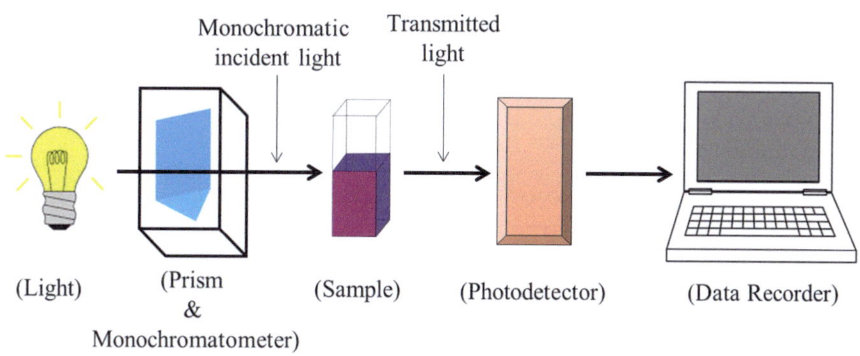

Fig. 16.1 Schematic presentation of a spectrophotometer and its parts

3.4 DNA Sample

Extracted DNA from reference as well as forensic samples can be used for the estimation of DNA quantity and quality. If you expect a higher quantity of DNA, dilute the extract in TE buffer prior to taking the absorbance of the sample. Various known concentrations of calf thymus DNA (CT DNA) prepared in TE buffer can be used as control DNA for the assessment of measurement of DNA concentration by UV–visible spectrophotometer.

4 Procedure

1. Turn on the power of spectrophotometer (Beckman DU64), and check that all accessories like printer are ready.
2. Turn on the light source (UV light) and allow it to warm for at least 5 min.
3. Select the absorbance mode using "ABS" key.
4. Press the "SCAN" key.
5. Go to "EDIT" and enter starting wavelength as "280 nm" and press "Enter."
6. Similarly, enter ending wavelength at "260 nm" and press "Enter."
7. Manage the speed of the sample scan at 750 nm/min by pressing the "STEP" key and scrolling through the options to find 750 nm/min followed by pressing of "Enter."
8. Set the upper limit at 2000 absorbance; press "Enter."
9. Set the lower limit at 0.000 absorbance; press "Enter." Now the starting wavelength will reappear.
10. The instrument now becomes ready for calibration against a control solution.
11. Place 200 μl of TE buffer in a quartz cuvette, and open the sample compartment lid of the instrument.
12. Carefully clean the cuvette using a Kimwipe® and place the cuvette containing TE buffer into the cuvette holder. Make sure that the quartz sides of the cuvette are in the path of light source.
13. Close the lid of the sample compartment.
14. Press "CALB."
15. The absorbance of TE buffer is being stored in the memory as background and "Bkg" will appear in the display.
16. Press "READ."
17. Wait till appearing of "Scan," which shows that the calibration is complete and the instrument is ready for sample analysis.

18. Open the sample compartment, remove cuvette, and discard 200 μl of TE buffer present in the cuvette onto the kimwipe.
19. Place 200 μl of solution containing DNA in EDTA in cuvette.
20. Place the cuvette in the sample holder.
21. Press "READ."
22. The absorbance of the sample between 260 and 280 nm will be measured and plotted as a graph.

5 Observation

For each DNA sample analyzed, note down the absorbance at 280 and 260 nm. Additionally, divide the absorbance at 260 nm with the absorbance at 280 nm to gather the information regarding the quality of DNA. To determine the value of DNA concentration multiply the value of absorbance with the constant, i.e., 50 μg/ml.

6 Result Table

Sample	DNA content	$OD_{260/280}$	Inference
Control			
Sample 1 (*liquid blood*)			
Sample 2 (*forensic sample*)			

OD optical density

7 Precautions

- Wear gloves while dealing with the reagents and plasticware.
- Do not touch the quartz size of the cuvette.
- Do not look at the light source directly.
- Clean the cuvette thoroughly before and after measurement.
- Use freshly diluted DNA samples for measurement of their concentrations.
- Always fill the cuvette to 3/4th of its capacity.
- Use low-retention microtips during preparation of DNA samples.
- Dispose the reagents and consumables carefully after the examination.

8 Troubleshooting

Problem	Tentative cause	Possible solutions
Inappropriate absorbance (OD) value	DNA concentration is too high	Dilute the sample in TE buffer and take the reading again
	Quartz side of the cuvette is not present in the line of light source	Change the orientation of the cuvette in the sample holder
	Instrument not working properly	Call the service person of the instrument
The initialization of the spectrophotometry showing >>>FAILED<<<<	Instrumental failure	Click on "Cancel button"; power off the instrument, wait for few seconds, and power it on again
	Electricity problem	
The machine showing various dialogues:	"Blank required"	Place the cuvette with TE filled in it and take the measurement by clicking on "BLANK" icon button
	"UV Lamp required"	Switch on the UV source light. Wait for few minutes till the UV light is warming up
	"Visible Lamp required"	Switch on the visible light source. Wait for few minutes till the visible light is warming up

Chapter 17

Quantification of DNA by Using qRT-PCR

Objective: To assess the quality and quantity of DNA isolated from blood and forensic samples using qRT-PCR technique.

1 Introduction

Careful qualitative and quantitative evaluation of DNA is an imperative for forensic DNA analysis. Most of the DNA samples of forensic origin contain a large number of PCR inhibitors, degraded samples, as well as various exogenous contaminants, which affect the downstream processing of DNA adversely. The commonly used DNA quantitation techniques such as gel electrophoresis method or UV–visible spectrophotometer technique are less accurate and time consuming or they consume a substantial amount of the precious sample. To overcome these problems and to increase the accuracy and precision of DNA quantitation method, advanced real-time quantitative PCR (qRT-PCR) technique has been discovered with an increased efficiency to analyze the challenged, complex-natured forensic DNA samples.

During forensic DNA typing, PCR is the must prerequisite to amplify the autosomal as well as other STR markers. This sensitive step requires an optimum input of DNA quantity to avoid/minimize artifacts in the results which may include stochastic amplification and/or oversized DNA profile. Thus, a real-time qPCR technique is preferred over other techniques using 5′-nuclease fluorogenic or TaqMan® assays. This technique has many advantages over the other techniques such as generation of a linear response proportional to DNA quantity and its occurrence in a closed system which minimizes the potential carryover of contaminants. The automation associated with this system is an added advantage by the use of automated fluorescence detection systems. Currently many human-specific q-RT PCR assays are available

H. R. Dash et al., *Principles and Practices of DNA Analysis: A Laboratory Manual for Forensic DNA Typing*, Springer Protocols Handbooks, https://doi.org/10.1007/978-1-0716-0274-4_17,

which target autosomal, Alu repeats, Y chromosome, and mtDNA targets. Multiplex techniques are currently being used for simultaneous detection of large and small human telomerase reverse transcriptase gene (hTERT), a synthetic internal positive control, and male DNA fragment to give rise to information on total human DNA quantity, male DNA vs. female DNA ratio, degradation index, as well as information on potential PCR inhibitor. Many commercially available kits are available (for details see Appendix B), out of which Quantifiler® Trio DNA Quantification Kit of Thermo Fisher Scientific has been described in this chapter.

2 Principle

Four 5′ nuclease assays are carried out simultaneously in Quantifiler® Trio DNA Quantification, i.e., human-specific short amplicon assay, human-specific large amplicon assay, human male DNA-specific assay, and assay for internal positive control (IPC). The kit works in TaqMan® chemistry, where each PCR primer of each target possesses specific dye-labeled probes and nonfluorescent quenchers as follows (Table 17.1):

The technique relies on the 5′ nuclease assay which takes place in each cycle of PCR process, which does not interfere with the end product formation (Fig. 17.1). During PCR process, the TaqMan® probe containing both the reporter dye and quencher compound binds to the complementary sequence of the target DNA in between the forward and reverse primer. In this step, the proximity between the reporter dye and quencher suppresses the fluorescence level of the reporter dye mostly by Förster-type energy transfer. During the extension of primer in PCR process, when the growing strand arrives at the point of attachment of TaqMan® probe due to action of Taq polymerase, the TaqMan® probe is cleaved due to the action of Taq polymerase and the reporter dye and quencher

Table 17.1
Targets and their associated dye and quenchers

Sl. no.	Target	Length of amplicon (bp)	Copy number	Dye	Quencher
1	Human-specific large autosomal	214	Multi-copy	ABY™	QSY™
2	Human-specific small autosomal	80	Multi-copy	VIC™	MGB
3	Human-specific male DNA	75	Multi-copy	FAM™	MGB
4	Internal PCR control	130	Synthetic DNA template	JUN™	QSY™

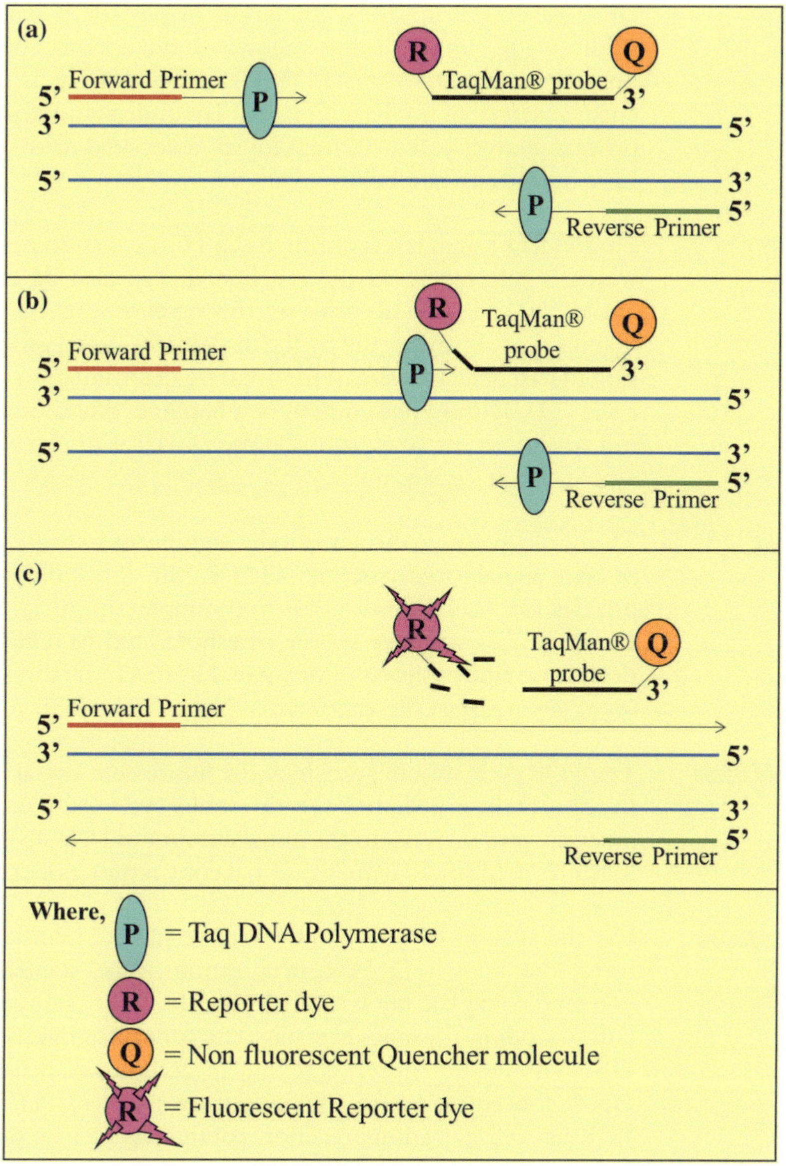

Fig. 17.1 Various steps of 5′ nuclease assay using TaqMan® probe for fluorescent detection of quantity of DNA using qRT-PCR. (**a**) Polymerization. (**b**) Strand displacement. (**c**) Polymerization complete

molecules are separated. This results in an increased fluorescence of the reporter dye. Thus, there occurs an increase in fluorescence level of the reporter dye only when the target DNA sequence complementary to the TaqMan® probe is amplified in PCR and hence the nonspecific amplifications cannot be detected.

The advantage of using qRT-PCR over other techniques is that it allows real-time detection of amplicon at each PCR cycle which results in the accurate detection of the amount of starting quantity DNA, monitoring the real-time progress of PCR process, dynamic range of detection, and elimination of post-PCR manipulations. Additionally, multiplexing allows simultaneous detection of human-specific DNA, male DNA, degraded DNA, as well as presence/absence of PCR inhibitors. To understand the qRT-PCR process better, a few terminologies associated with the process have been described below.

2.1 Baseline

Fluorescent signal levels during 3–15 PCR cycles do not vary much which is referred to as the baseline. It can also be called as the "noise." During PCR process, the baseline can be determined manually or automatically by the instrument. But a proper designation of baseline gives rise to the accurate determination of threshold cycle (C_t). During designation of a baseline, one should be careful not to choose the background noise as well as not to eliminate the amplification signals.

2.2 Threshold

Threshold signifies the statistically significant increase in the fluorescent signal over baseline. Hence, the threshold level distinguishes the background noise from amplification signal. Most of the qRT-PCR software set the threshold level to tenfold standard deviation value of the baseline value. However, the threshold can be set at any point of the exponential phase of the PCR process.

2.3 Threshold Cycle (C_t)

The PCR cycle number at which the fluorescent signal level crosses the threshold is called as the threshold cycle (C_t). As C_t value is inversely proportional to the initial amount of DNA in the reaction, it can be used to calculate the initial copy number of DNA present.

2.4 Standard Curve

For the analysis of qRT-PCR results, a standard curve is required using the known DNA concentrations. The standard curve is prepared using the known DNA concentrations at *x*-axis and their corresponding C_t values at *y*-axis. The concentrations of the known DNA should be chosen carefully to accommodate the expected DNA concentration of the unknown samples. Performance of PCR as well as various reaction parameters can be derived from the standard curve such as slope, *y*-intercept, and correlation coefficient.

2.5 Correlation Coefficient (R^2)

Correlation coefficient describes the authenticity of the standard curve. The ideal value of the R^2 is 1; however, the calculated value of 0.999 is considered to be excellent for the use of the standard curve in result interpretation.

2.6 y Intercept

The detection limit or sensitivity of the reaction is determined by the *y*-intercept. This result corresponds to the expected C_t value if the lowest amount of DNA present on the *x*-axis gives rise to a statistically significant fluorescent signal. Most of the qRT-PCR applications are capable of detecting a DNA amounting to 2^{-10} copy numbers, which is considered as the lowest target level.

2.7 Exponential Phase

At the early part of the exponential phase of amplification, all the reagents such as DNA polymerase and primers are present in excess amount. Hence, it is recommended to quantify the DNA concentration at the early exponential phase of the reaction to get a better, accurate, and sensitive result.

2.8 Slope

The efficiency of a q-RT PCR reaction is measured by the slope of the log-linear phase. A slope of −3.32 corresponds to 100% efficiency of a q-RT PCR process. The slope value between −3.58 and −3.10 can be considered as a good-quality PCR process.

2.9 Efficiency

$$\text{Efficiency} = 10^{(-1/\text{slope})} - 1$$

100% Efficiency of a PCR process is the doubling of the amount of DNA after each PCR cycle in the exponential phase. Based upon the above equation, the ideal efficiency of the reaction can only happen when the slope value will be −3.32. A reaction is considered to be good at the efficiency range of 90–110%. A reduction in PCR efficiency below 90% occurs when nonoptimal reagent concentration and poor enzyme quality are used.

2.10 Dynamic Range

The range over which the increase in template DNA concentration results in the amplification product is called as the dynamic range. Ideal dynamic range for genomic DNA samples should be in the 3–4 log range.

3 Reagents/Materials Required and Their Role

3.1 Quantifiler™ THP PCR Reaction Mix

It contains dNTPs, buffer, enzyme, Mustang Purple™, and stabilizers. Mustang Purple™ is used as the passive reference dye. Stabilizers are generally used in a PCR mix to increase the stability of the PCR components and to increase the efficacy of a PCR process. The common PCR stabilizer may include trehalose sugars or glycerol. Upon receipt of this component at −25 to −15 °C temperature, it can be stored away from direct light at 2–8 °C till further use.

3.2 Quantifiler™ Trio Primer Mix

It contains the target-specific primers; probes labeled with ABY™, JUN™, VIC™, and FAM™ dyes; as well as template DNA for internal PCR control (IPC). As this is a multiplex process, the added primers correspond to the DNA fragments specific to human-specific large fragment, human-specific small fragment, and human-specific male DNA and for the internal PCR control and their corresponding probes the dyes are ABY™, VIC™, FAM™, and JUN™. Upon receipt of this component at −25 to −15 °C temperature, it can be stored away from direct light at 2–8 °C till further use.

3.3 Quantifiler™ THP DNA Dilution Buffer

The dilution buffer mostly contains the low TE buffer to be used for preparing various dilutions of the DNA standards. Upon receipt of this component at −25 to −15 °C temperature, it can be stored at 2–8 °C till further use.

3.4 Quantifiler™ THP DNA Standard

It contains genomic DNA standard at a concentration of 100 ng/μ l. This stock concentration is diluted using Quantifiler™ THP DNA Dilution Buffer to obtain various dilutions to be used for standard curve preparation. Upon receipt of this component at −25 to −15 °C temperature, it can be stored at 2–8 °C for further use.

4 Procedure

Preparation of Standards

1. Label five 1.5 ml microcentrifuge tubes as Std. 1, Std. 2, Std. 3, Std. 4, and Std. 5.
2. Dispense 10 μl Quantifiler™ THP DNA dilution buffer to the tube marked as Std. 1 and 90 μl of the same to the tubes Std. 2 to Std. 5.
3. Add 10 μl of Quantifiler™ THP DNA Standard to the buffer containing tube marked as Std. 1. As the concentration of Quantifiler™ THP DNA Standard is 100 ng/μl, the final concentration becomes 50.000 ng/μl in Std. 1.
4. Vortex Std. 1 carefully for 3–5 s, and dispense 10 μl from Std. 1 to the buffer-containing tube marked as Std. 2 using a fresh microtip. The final concentration of DNA in Std. 2 becomes 5.000 ng/μl.
5. Mix Std. 2 carefully by vortex for 3–5 s and dispense 10 μl from Std. 2 to the buffer-containing tube marked as Std. 3 using a fresh microtip. The final concentration of DNA in Std. 3 becomes 0.500 ng/μl.
6. Mix Std. 3 carefully by vortex for 3–5 s and dispense 10 μl from Std. 3 to the buffer-containing tube marked as Std. 4 using a fresh microtip. The final concentration of DNA in Std. 4 becomes 0.050 ng/μl.
7. Mix Std. 4 carefully by vortex for 3–5 s and dispense 10 μl from Std. 4 to the buffer-containing tube marked as Std. 5 using a fresh microtip. The final concentration of DNA in Std. 5 becomes 0.005 ng/μl.

PCR Setup

8. Calculate the requirement of each component of the PCR process using the following table as per your requirement/ number of samples (Table 17.2).

Table 17.2
Amounts of reagents required to prepare master mixture for q-RT PCR

qRT-PCR components	Volume per reaction (μl)	No. of reactions including standards	Total volume
Quantifiler™ THP PCR Reaction Mix	8	A	8A
Quantifiler™ Trio Primer Mix	10	A	10A
Total volume			= (8A + 10A)

9. Thaw the Quantifiler™ THP PCR reaction mix and Quantifiler™ Trio primer mix, vortex for 3–5 s, and centrifuge briefly before using.
10. Take out the requisite quantity of reaction mix and primer mix to a fresh microcentrifuge tube using separate fresh microtips.
11. Mix the reagents of the tubes containing master mixture by vortex of 3–5 s.
12. Pipette out 10 μl of master mixture into each well of a 96-well plate or PCR tube.
13. Add 2 μl of standards, +ve/−ve control, and unknown samples to the designated wells/tubes.
14. Use the optical adhesive cover to seal the 96-well plate or the optical 8-cap strip for the strip tube.
15. Inspect each well of the plate/each tube for the presence of any air bubbles. If present, tap each well with fingertip to remove the same.
16. Rotate the plate in a centrifuge at 3000 rpm for 30 s. Now the plate is ready to be loaded in the machine.

Experimental Setup

17. Switch on the computer by entering the required user name and password.
18. Press the power button of 7500 Real Time PCR System present at its lower right front.
19. Wait till the establishment of communication between the computer and the instrument when the green power indicator is lit.
20. Go to Start Programs → Applied Biosystems → HID Real-Time PCR Analysis software → HID Real-Time PCR analysis software v1.2.
21. Log in to the software using the proper "user name" and "password" or log in as "guest".

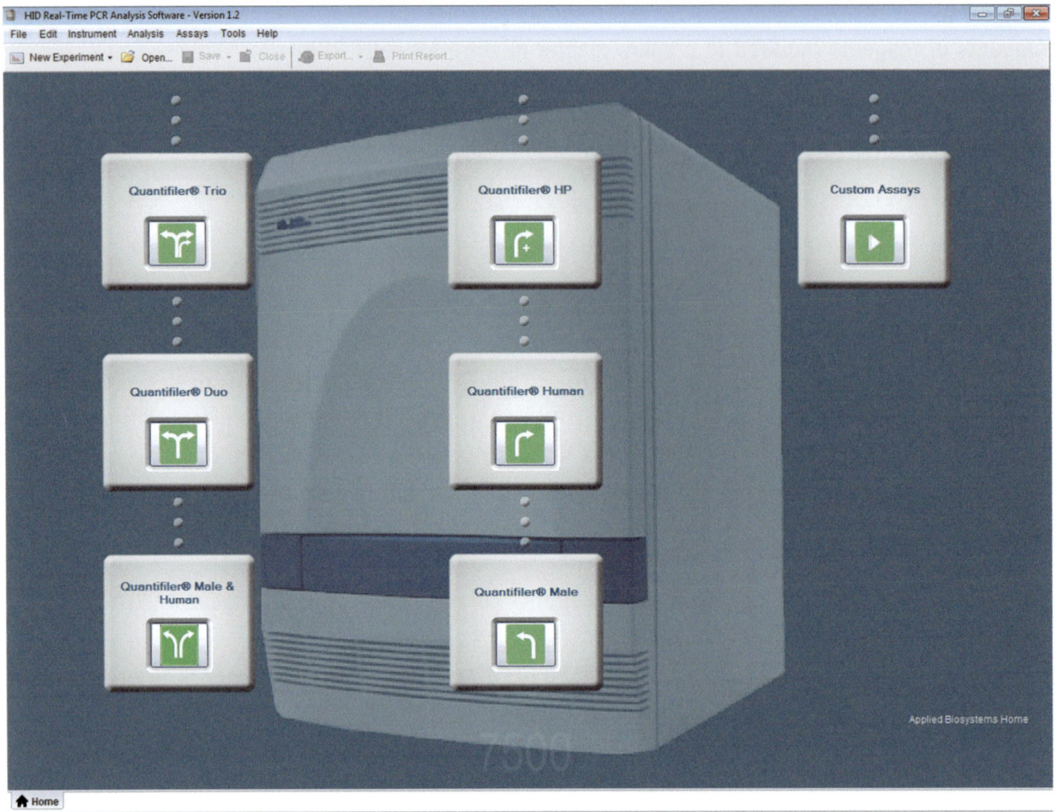

Fig. 17.2 Home screen of the software shows all the available kits for HID analysis. You have to select the icon as per your use

22. After logging in to the software, the home screen will appear as Fig. 17.2. Select the suitable application, i.e., Quantifiler™ Trio for the current application.
23. After selecting the suitable application, the experiment properties will appear in the screen (Fig. 17.3), where the name of the experiment is filled. Rest credentials are autofilled by the software.
24. Click on "Plate Setup." Targets will be found to be prefilled as per the selected assay.
25. Go to define samples and click on "Add New Sample" (Fig. 17.4a, b). Fill the name of the sample. Repeat the process for the samples as per your requirement.
26. Go to "Assign Targets and Samples" (Fig. 17.5). Click on the well to select the sample which is present on that well. On the left side of the plate layout select the desired sample in "Assign sample(s) to the selected wells" for which you just selected the well. Each sample is to be selected by default.

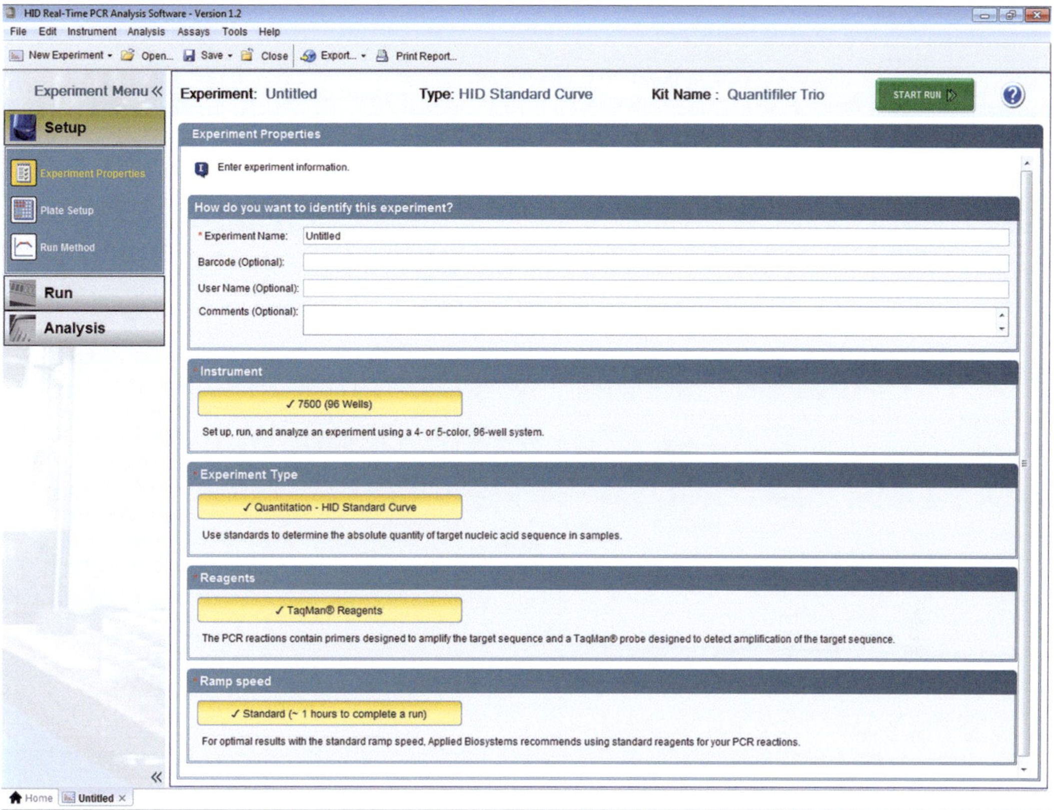

Fig. 17.3 “Experimental properties” showing credential to be filled, i.e., “Experiment name”

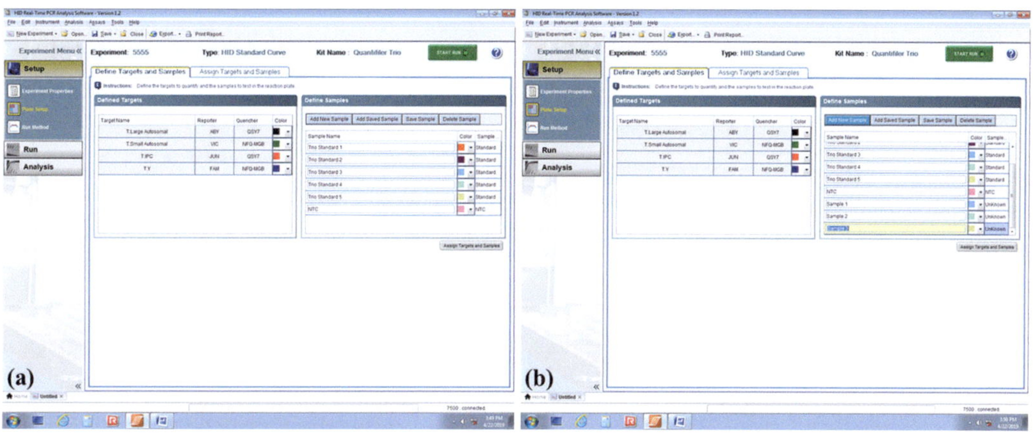

Fig. 17.4 Define samples by clicking on the “Add new samples” icon as per your requirement and fill the sample name. (**a**) Default icons. (**b**) Addition of samples to be analyzed

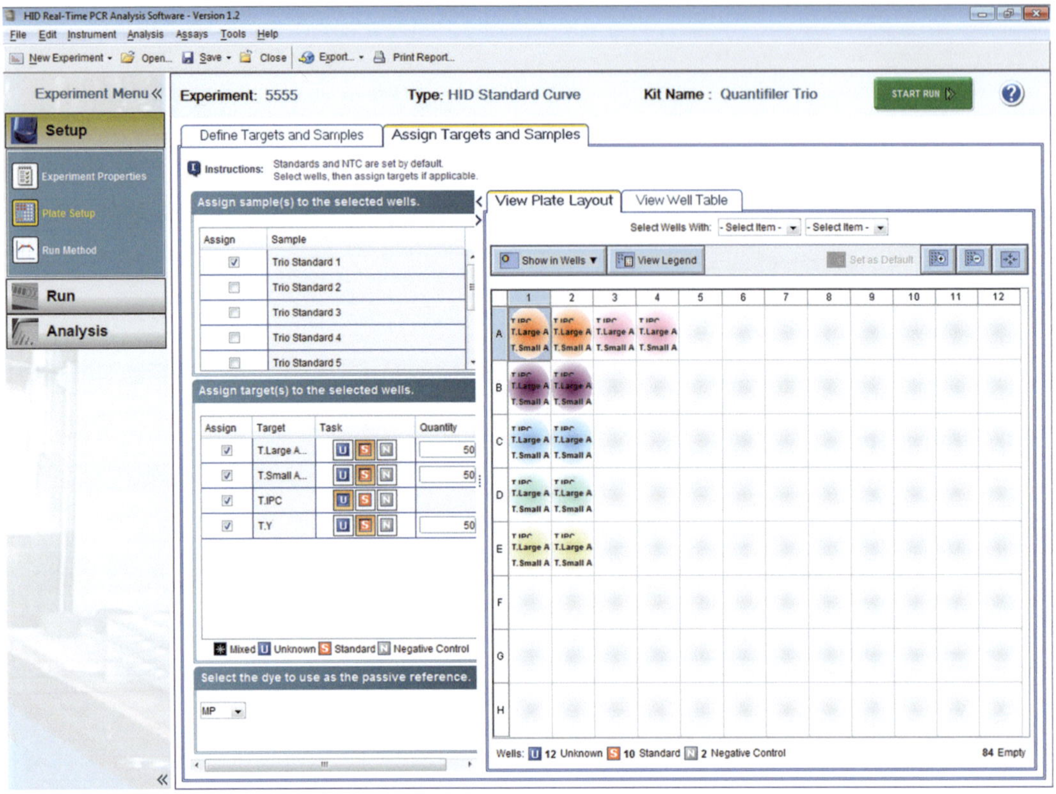

Fig. 17.5 Assign the samples and their corresponding targets in the well

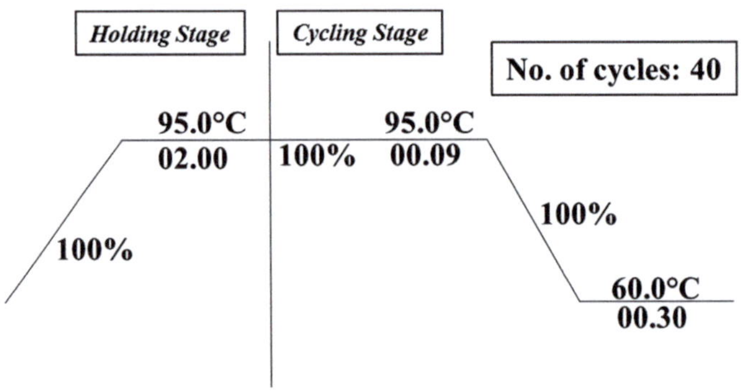

Fig. 17.6 Display of Run Method for the run protocol of Quantifiler™ Trio kit

27. Select the "Run Method." The run protocol for Quantifiler™ Trio kit has been set earlier which will be displayed (Fig. 17.6).
28. Click on "save" to save the protocol.

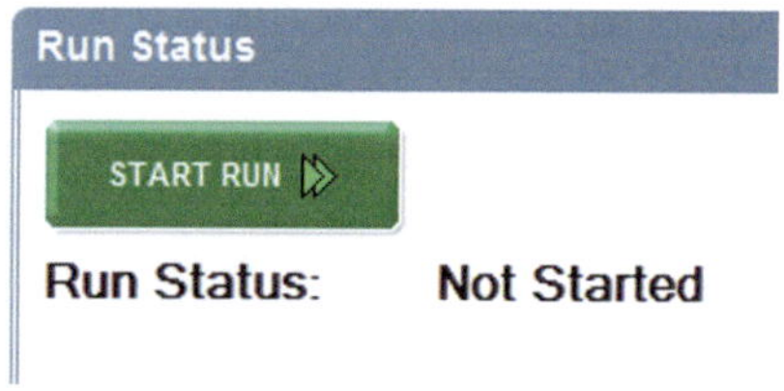

Fig. 17.7 Click on "Start Run" after loading the plate in the instrument

Loading of the Plate in the Instrument

29. Press the "door" of the 7500 Real Time PCR instrument to open the tray/plate holder.
30. Keep the plate on the plate holder with correctly aligned manner and close the door.
31. Open the software, open the protocol that has been saved earlier, and click on "Start Run" (Fig. 17.7).

Analysis of Result

32. After completion of run, check for "Amplification Plot," "Standard Curve," "Multicomponent Plot," "Raw Data Plot," "QC Summary," and "Multiple Plots view" in the section "Analysis" (Fig. 17.8).
33. To export the results, in the experiment menu, select "Analysis" and "view plate layout," select the wells to export and select data to export, and click on "Start Export" (Fig. 17.9).

5 Observation

To access the quality of qRT-PCR result, the following parameters should be taken care of:

- Amplification plot for small autosomal, large autosomal, and internal PCR control should be present above C_t value.
- Amplification plot should not be present for NTC, i.e., no template control.
- All the replicates of the sample/standard should show the same or near-similar results.
- Standard curve should be linear when 5 pg/μl to 100 ng/μl of samples are used.

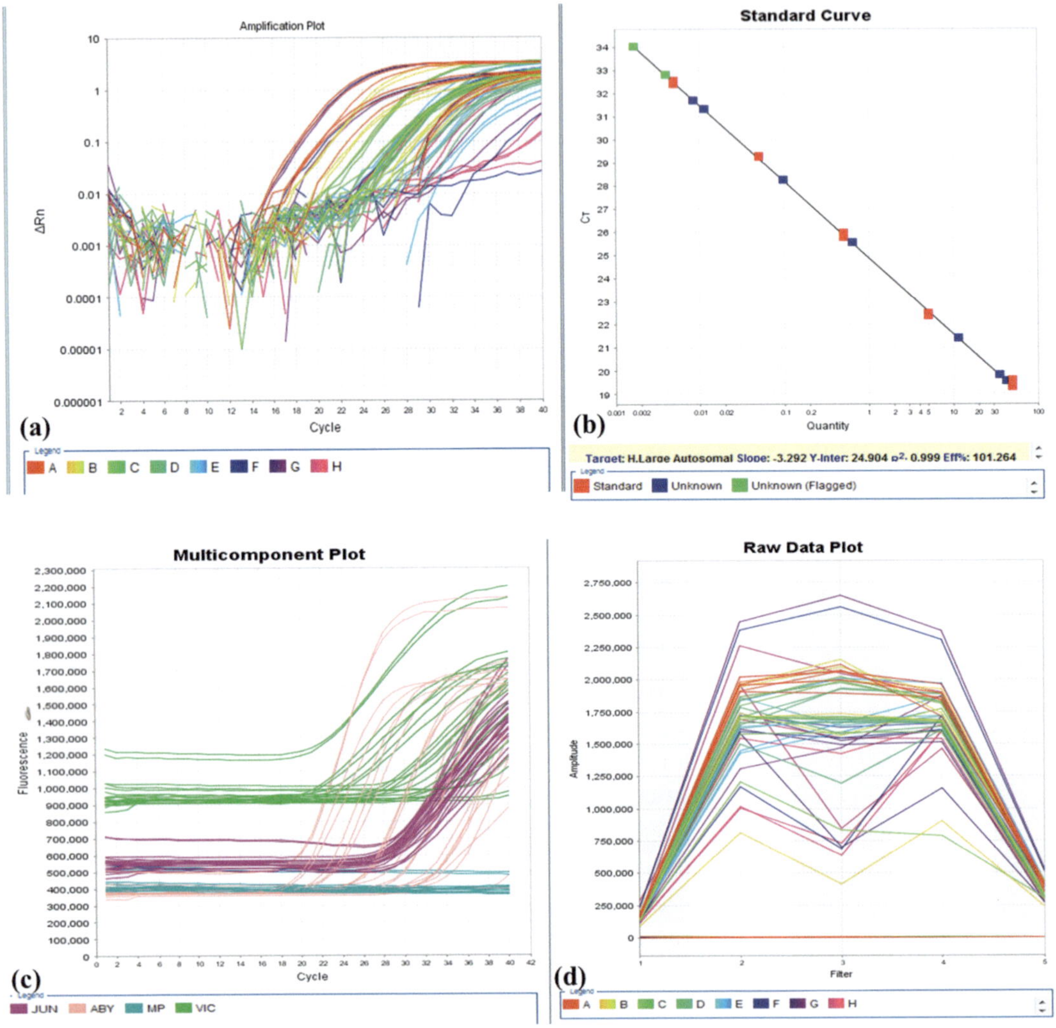

Fig. 17.8 Various parameters for accessing the qRT-PCR results, i.e., (**a**) amplification plot, (**b**) standard curve, (**c**) multicomponent plot, and (**d**) raw data plot

- The R^2 value should be ≥ 0.999.
- The slope should be close to -3.3.
- An increase in template DNA concentration may increase the C_t value of the internal PCR control due to competition of PCR components.
- A negative result is observed in case when no human DNA is present. In such cases, amplification signals from small and large autosomal and Y-specific targets are not present, but normal amplification from internal PCR control is observed.

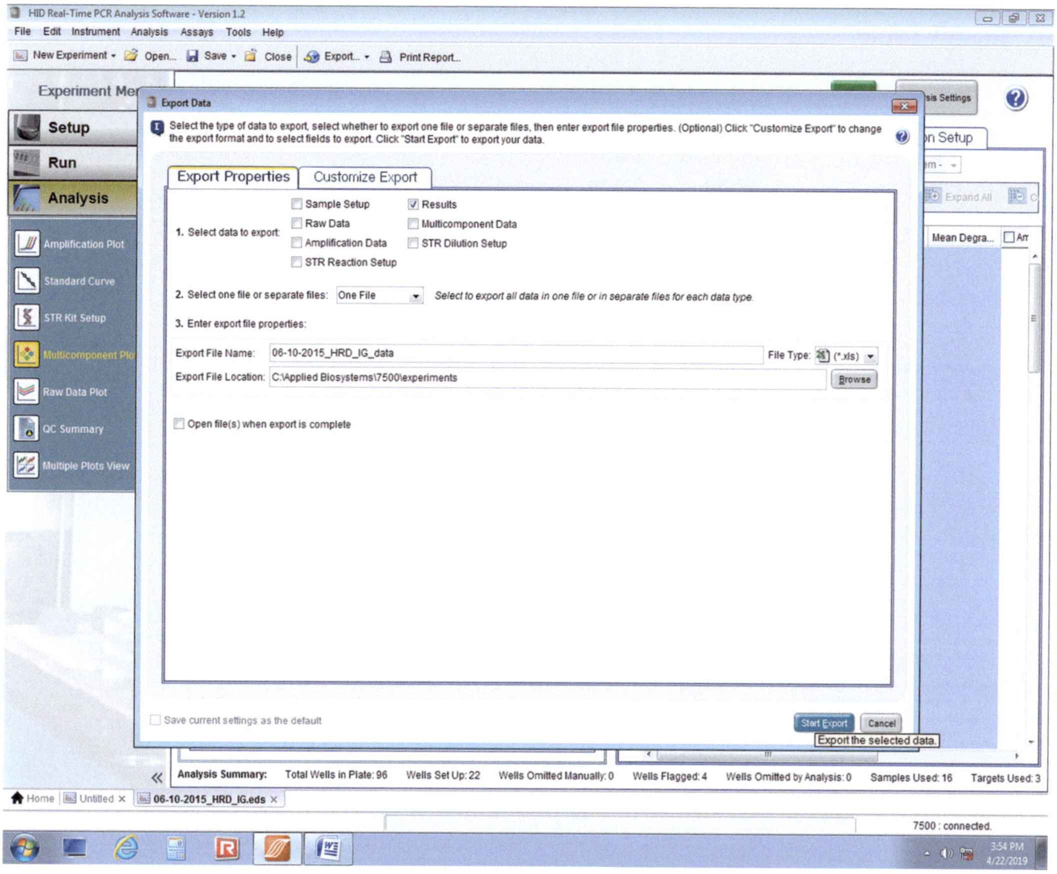

Fig. 17.9 Export and save qRT-PCR results to the folder as per your requirement

- A complete failure in PCR process results in the absence of results for all targets, i.e., small and large autosomal DNA, Y-specific target, as well as internal PCR control.
- Low or weak amplification of IPC is an indication of the presence of potential PCR inhibitors in the reaction.
- The degradation index is calculated as small autosomal target DNA concentration (ng/μl)/large autosomal target DNA concentration (ng/μl). A degradation index of <1 signifies a good quality of sample, whereas the value in between 1 and 10 indicates that the sample is moderately degraded. However, when the value becomes >10 it indicates a highly degraded DNA sample.

6 Result Table

Sample name	Target name	C_t	Quantity	M:F ratio	Degradation index	IPCC T	NTCC T
Std. 1	IPC						
	Large autosomal						
	Small autosomal						
	Y						
Std. 2							
Std. 3							
Std. 4							
Std. 5							
Blood sample							
Bone							
Vaginal swab							
Hair							
Teeth							

7 Precautions

- Wear gloves while dealing with the reagents and plasticware.
- Do not store the reagents under a light source.
- Always use freshly prepared dilutions of the standards for the preparation of standard curve.
- Always use filtered microtips to prevent any possible contaminations.
- First switch on the computer and then switch on the instrument followed by switching on the software.
- Always use the standards in triplicates to obtain results with higher R^2 value.
- As qRT-PCR is a very sensitive technique, always obey the three-room rule for extraction, setup, and qRT-PCR setup.

- Use freshly calibrated pipettes for pipetting of accurate small volumes. Small pipetting error may give rise to ambiguity in results.
- Always use an optimized protocol. Use the manufacturer's guidelines properly.

8 Troubleshooting

Problem	Tentative cause	Possible solutions
Slope of the standard curve is outside the normal range or R^2 value is <0.98	Incorrect quantity of standards has been applied	Update with correct values and reanalyze
	A stored dilution of the standards was used	Prepare fresh standards and rerun the samples
	Standards were not loaded	Exclude a failed/flagged standard in standard curve analysis, and reanalyze the samples
Replicates of the samples and/or standards showing inconsistent ΔRn and C_t results	Inappropriate amount of Quantifiler™ THP PCR reaction mix was added	Check the result in multicomponent plot
		Compare the volume of the reaction mix in the well with discrepancy in result with the nearby wells
High C_t value and low ΔRn value	Presence of high quantity of PCR inhibitors	Take a chance by diluting the samples before multiplex PCR or purify the sample again and reanalyze
Positive result observed in negative control samples	Improper baseline selection detects small amount of fluorescence signals as amplifications	Change the baseline manually
	Contamination to consumables and/or reagents	Rerun with fresh batch of kits, consumables, and plasticware

Chapter 18

Fluorescence-Based Instant Quantification of DNA

Objective: To quantify DNA isolated from blood and forensic samples using fluorometric technique by Qubit™ Fluorometer.

1 Introduction

Measurement of DNA concentration by fluorometric technique relies on the use of dyes that selectively bind to DNA. Any of the DNA extraction protocol does not eliminate RNA completely. Though many of the protocols recommend the use of RNase, it can only help in reducing the chance of errors. Due to the presence of RNA contamination in extracted DNA, UV–visible spectrophotometer-based techniques do not yield an accurate result of DNA concentration. In this regard, fluorescence-based technique is more useful as it allows the measurement of double-stranded DNA-dye complex selectively. On binding with the double-stranded DNA, the dye undergoes a conformational change and subsequently a fluorescent energy is emitted which is detected by the instrument.

The examples of such dyes include ethidium bromide, SYBR® Green, Hoechst (bis-benzimide) dyes, and PicoGreen®. In this case, the amount of fluorescence generated by the dyes is directly proportional to the concentration of DNA in the sample. Thus, the concentration of DNA is measured indirectly by estimating the fluorescence level of the dye in the solution. There are many advantages of the use of fluorescence-based DNA quantification technique. The main advantage includes the selective binding of the dyes to DNA and nonbinding to other possible co-extracts during DNA isolation, i.e., protein or RNA. Additionally, the strong affinity between the DNA and the dye allows detection of DNA at a very low concentration. The detection of low signal generated due to trace amount of DNA sample makes this technique more sensitive

H. R. Dash et al., *Principles and Practices of DNA Analysis: A Laboratory Manual for Forensic DNA Typing*, Springer Protocols Handbooks, https://doi.org/10.1007/978-1-0716-0274-4_18, © Springer Science+Business Media, LLC, part of Springer Nature 2020

than spectroscopic technique with the highest degree of accuracy and affinity. Some of the disadvantages of this technique include the cost of the fluorescent dyes and the associated health risks of the fluorescent molecules.

2 Principle

Many fluorescent dyes are available which specifically bind to double-stranded DNA specifically. Some of these dyes include ethidium bromide, SYBR® Green, Hoechst (bis-benzimide) dyes, and PicoGreen®. After binding to the double-stranded DNA, these dyes undergo conformational changes which lead to the generation of fluorescent energy to be detected by an instrument. In this case, the amount of fluorescence generated is directly proportional to the DNA present in the solution. First a standard curve/calibration curve is prepared (Fig. 18.1) by measuring the fluorescence level through a fluorometer corresponding to various concentrations of DNA. Finally, the fluorescence level of an unknown sample is measured and its corresponding DNA concentration is calculated based on the calibration curve.

For better understanding a commonly used dye Hoechst 33258 is described below. Hoechst 33258 is a common nucleic acid stain that binds to double-stranded DNA categorically and emits blue fluorescence light upon binding. The dye is excited by the wavelength 352 nm, whereas its emission maximum is at the wavelength of 461 nm. This dye can be excited by using mercury or xenon lamp or using UV source. The dye is soluble in both water and organic solvents such as DMSO or dimethylformamide.

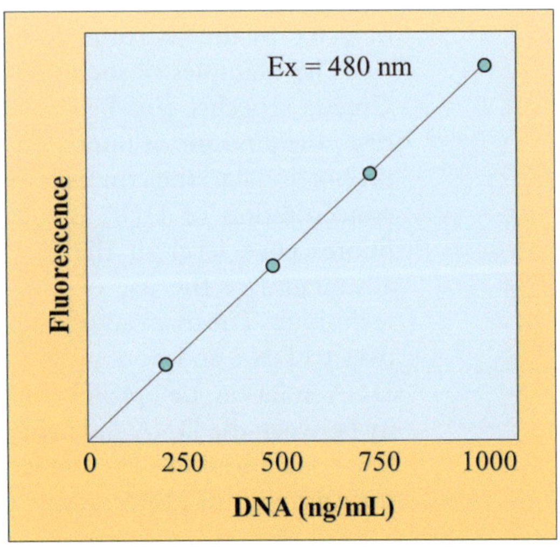

Fig. 18.1 Calibration curve of control DNA (calf thymus DNA) when PicoGreen® is used as the double-stranded DNA quantitation reagent

Hoechst 33258 binds to the minor groove of double-stranded DNA. DNA sequences rich in adenine and thymine have a preferential high binding with the dye. Another advantage of this dye is that it is permeable to cell membrane and can bind to the DNA of both live and fixed cells. To prepare 1 mg/ml Hoechst 33258 solution, add 10 mg of the dye powder to 10 ml of distilled water, and do not filter. The prepared dye solution can be used within 6 months by proper storage at 4 °C.

Though many kits are available in the market, Qubit™ 1× dsDNA HS Assay Kit has been described in this chapter for fluorescence-based DNA quantification using Qubit™ 4 Fluorometer model.

3 Reagents/Materials Required and Their Role

3.1 Qubit™ 1× dsDNA HS Working Solution (Component A)

This component of the kit contains the fluorescent dye at 1× concentration. It should be stored at 2–8 °C under protection from light. The dye present in component A is highly selective for double-stranded DNA. It can work for a wide range of DNA concentrations, i.e., 10 pg/μl to 100 ng/μl, or 0.1–100 ng. The fluorescent signal generated during this reaction is highly stable till 3 h when it is protected from exposure to light. Common contaminants such as salts, free nucleotides, solvents, detergents, or proteins do not affect this assay of measurement of DNA concentration.

3.2 Qubit™ 1× dsDNA HS Standards #1 and #2 (Components B and C)

Both component B and component C are required for calibration of the instrument prior to measurement of unknown DNA concentration. Standard #1 and Standard #2 contain 0 and 10 ng/μl of DNA concentration in TE buffer, respectively. They should be stored at 2–8 °C.

3.3 TE Buffer

TE buffer is generally prepared by mixing 50 mM Tris and 50 mM EDTA at pH 8.0. Tris, the major constituent of TE buffer, acts as a common pH buffer. Additionally, EDTA chelates cations like Mg^{2+}. Hence, TE buffer is helpful to solubilize DNA by protecting it from degradation.

3.4 DNA Sample

Extracted DNA from reference as well as forensic samples can be used for the estimation of DNA. If you expect a higher quantity of DNA, dilute the extract in TE buffer prior to taking the absorbance of the sample. Various known concentrations of calf thymus DNA (CT DNA) prepared in TE buffer can be used as control DNA for assessment of measurement of DNA concentration by fluorometric technique.

4 Procedure

Preparation of Standards and Samples

1. Label as many number of 500 μl microcentrifuge tubes as your number of samples on their lids and two additional tubes for two standards.
2. Add 10 μl of both the standards and samples to the corresponding tubes.
3. Add Qubit™ 1× dsDNA 1× buffer to each tube to make up the final volume to 200 μl.
4. Mix both the solutions by vortexing for 3–5 s.
5. Incubate the tubes at room temperature for 2 min.

Measurement of DNA Concentration

6. Switch on the Qubit™ 4 Fluorometer.
7. Press "DNA," and then select the assay type as "1× dsDNA HS."
8. Then the display will show "Read Standards." Press on the same to proceed.
9. Insert the tube containing Standard #1 in the sample chamber followed by closing of the lid and press "Read Standard."
10. When reading is complete (which takes around 3 s), remove Standard #1.
11. Insert the tube containing Standard #2 in the sample chamber followed by closing of the lid and press "Read Standard."
12. When reading is complete (which takes around 3 s), remove Standard #2.
13. Press "Run Samples."
14. Adjust the sample volume between 1 and 20 μl.
15. Select the units for output sample concentration from the drop-down menu, i.e., ng/μl, ng/ml, μg/μl, μg/ml, or mg/ml.
16. Insert the tube containing unknown DNA concentration to the sample chamber, close the lid, and press "Read tube."
17. Wait till reading of the sample (which takes around 3 s) is over.
18. Remove the sample and repeat for any other samples.

5 Observation

The DNA quantity is displayed on the screen. The displayed value of DNA concentration is related to the concentration of the sample after dilution in the assay tube. The original concentration of the sample is calculated automatically in Qubit™ 3 and 4 Fluorometers, whereas for Qubit™ 2.0, the original DNA concentration should be calculated manually as follows:

$$\text{Concentration of the sample} = \text{QF value} \times 200/x$$

where QF = value given by the Qubit™ 2.0 Fluorometer and x = amount of sample added to the assay tube (in μl).

6 Result Table

Sample	DNA concentration
Control	
Sample 1 (*liquid blood*)	
Sample 2 (*bone powder*)	

7 Precautions

- Follow the manufacturer's guidelines in each step properly.
- Wear gloves while dealing with the reagents and plasticware.
- Avoid using an expired kit.
- Do not incubate the samples more than the recommended period of time.
- Do not hold the assay tubes in your hand because it warms the solution and may give rise to a different reading.
- Avoid keeping the sample directly in light.
- Component A may contain potential mutagens; hence handle it with care and dispose it with proper regulation.
- Store all the kit components at the recommended temperature condition.
- Never keep/install the Qubit™ instrument in a wet/humid place.
- Do not use wet hands to operate the Qubit™ instrument.
- Carefully operate the Qubit™ instrument so that the sample does not spill in the sample chamber.
- Make sure that the tube is clean and dry from outside while taking readings.
- Carefully handle the samples prior to taking the readings so that bubbles are not present in the sample.

8 Troubleshooting

Problem	Tentative cause	Possible solutions
In appropriate reading	The fluorescence dye is degraded due to inappropriate storage	Use the fresh reagents
	Bubbles present in the sample	Slightly tap on the tube wall or centrifuge briefly to remove the bubbles
	Outer surface of the tube is wet	Wipe the outer side of the tube using a Kimwipe
	Sample chamber is wet	Wipe the sample chamber using a Kimwipe
	Improper binding of the dye with the sample	Use fresh regents
		Follow recommended incubation conditions of the samples and standards
High reading of the sample/ out of range	Sample contains more DNA	Dilute the sample and retake the reading
	Lid of the instrument not closed while taking reading	Make sure to close the lid of the instrument while taking reading
	A high/low temperature of the sample while taking readings	Make sure that the assay is performed at room temperature/the recommended temperature
Low reading of the sample	Sample contains less DNA	Use a sample with lower dilution
	Standard and working solutions are not prepared correctly	Use recommended conditions to prepare the standard and working solutions and reread the samples
	Total volume is less	Make sure that the total volume of working solution + standards/samples is 200 μl
	Degraded component A	Protect the component A from light, use a fresh set of reagents, and retake the reading
	A high/low temperature of the sample while taking readings	Make sure that the assay is performed at room temperature/the recommended temperature

Part V

Amplification of Various STR Markers Using Multiplex PCR

Chapter 19

Amplification of Autosomal STR Markers by Multiplex PCR

Objective: To amplify the autosomal STR markers using multiplex PCR technique.

1 Introduction

Human identification and individualization are key in criminal justice system. Discovery of DNA fingerprinting technique has given rise to molecular level of identification with the highest degree of fidelity. The technology has provided results for many types of criminal and civil cases such as identification, murder, paternity dispute, as well as identification of sexual assault culprits. The beauty of this technique is that besides convicting the offenders, it can exonerate the innocence.

Human genome is comprised of 23 pairs of chromosomes, which include 22 pairs of autosomes and 1 pair of sex chromosome. Half of the autosomes are inherited from both the parents. The genomic complexity of an individual's autosomes is a result of chromosomal recombination and hence it is considered that no two individuals' genome is similar except the homozygous twins. This can be explored by autosomal STR analysis. Another well-known fact is that the two human beings share 99.9% of their genome and the genetic variation exists only in 0.1% of the genome. It was further found that the 0.1% variation occurs in the noncoding region of the genome. The interpersonal variation of the genome can be analyzed by determining the number of repeat units of the STRs, which are a key component in the noncoding region of the genome.

Short tandem repeats (STRs), also known as the microsatellites, are the short sequences of 2–7-nucleotide lengths which are repeated tandemly several times. Though many STR markers are present in

H. R. Dash et al., *Principles and Practices of DNA Analysis: A Laboratory Manual for Forensic DNA Typing*, Springer Protocols Handbooks, https://doi.org/10.1007/978-1-0716-0274-4_19, © Springer Science+Business Media, LLC, part of Springer Nature 2020

human genome, a few set of markers are being analyzed for forensic DNA purpose. To generate a coherent DNA profile, a common set of 20 markers are recommended to be analyzed called as the expanded CODIS markers which include CSF1PO, D3S1358, D5S818, D7S820, D8S1179, D13S317, D16S539, D18S51, D21S11, FGA, TH01, TPOX, vWA, D1S1656, D2S441, D2S1338, D10S1248, D12S391, D19S433, and D22S1045. Thus, multiplex PCR is a key step in forensic DNA analysis, where primers for all the STR markers are used to amplify the said STR loci simultaneously. As biological materials available in forensic samples are very less, PCR process plays a useful role in recovering genetic information from these samples.

The product size of a multiplex PCR is around 100–500 bp which is separated based on its size as well as fluorescent signals by capillary electrophoresis. Multiplex PCR process enables a high discrimination power of the test with a low input of DNA concentration, i.e., around 1 ng. Though many autosomal STR kits are available in the market (list of which has been given in Appendix B), amplification of autosomal STR markers using PowerPlex® Fusion 6C System has been described in this chapter.

2 Principle

During amplification of multiple targets in a single PCR setup multiplexing technique is applied. It varies from the conventional PCR by the use of multiple set of primers in a single reaction. In multiplex PCR process, designing of primers plays a useful role as all the primers should be of appropriate length, and should have similar melting temperature (T_m), greater specificity, as well as absence of any by-product formation such as formation of primer dimer. The greatest challenge in developing a multiplex reaction is that the primers should not react with each other. Additionally, it should contain homogenous molecules and a balanced primer concentration so that each amplicon is present in equal amounts.

Old-generation STR typing kits take around 2–3 h for completing the multiplex PCR process. Gradual advancements in technologies have reduced the time required for the same as well as improved the throughput of DNA typing. Some rapid polymerases are available in the market such as PyroStart (Fermentas, Glen Burnie, MD) and SpeedSTAR (Takara Bio USA, Madison, WI) which can be used to decrease the time requirement for multiplex PCR process. In this process, the thermal cycling protocol used for the amplification of genetic markers using PowerPlex® Fusion 6C System is as follows:

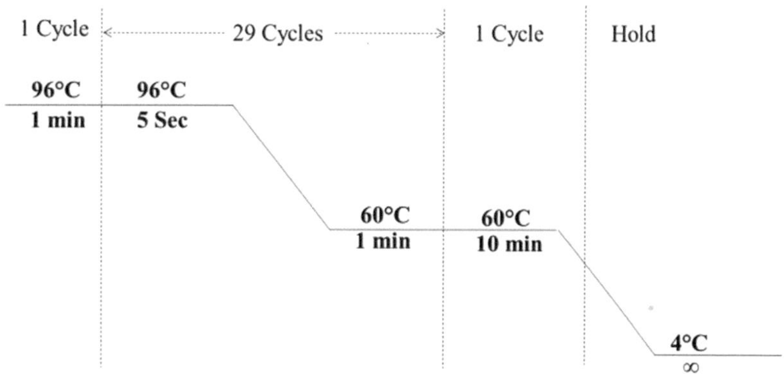

PowerPlex® Fusion 6C System is a six-dye multiplex system which allows co-amplification of 27 genetic markers for their fluorescent detection. The kit includes all 20 CODIS markers, i.e., CSF1PO, FGA, TH01, TPOX, vWA, D1S1656, D2S1338, D2S441, D3S1358, D5S818, D7S820, D8S1179, D10S1248, D12S391, D13S317, D16S539, D18S51, D19S433, D21S11, and D22S1045. Along with this, two sex-determining markers, i.e., amelogenin and DYS391, are also present in this kit. Penta D, Penta E, and SE33 have been included in this kit to increase the discrimination power. Additionally, two rapidly mutating Y-STR markers are also present in this kit as DYS570 and DYS576. The set of markers present in this kit satisfy both the expanded Combined DNA Index System (CODIS) and European Standard Set (ESS) recommendations.

The human specificity of the kit has been examined by amplifying unique exogenous DNA which includes 32 microbial sources, 19 non-primate vertebrate sources, and 4 unique primate sources, out of which none of the microbial sources yielded significant amplification. Though the kit showed some amplification for certain non-primates and primates, it was sufficient to entrust the human specificity of PowerPlex® Fusion 6C System. The sensitivity of the kit has been calculated to be as low as 250 pg. Additionally, the kit is highly useful for forensic applications as it works fine with common potential inhibitors such as hematin, humic acid, tannic acid, and EDTA. The precision study of the kit shows its usefulness on 3500, 3500xL, 3130, and 3130*xl* genetic analyzers. Experiments using this kit for mixtures, low template concentration, and degraded DNA have showed its usefulness in real-time forensic case work samples.

3 Reagents/Materials Required and Their Role

3.1 PowerPlex® Fusion 6C 5× Master Mix

Master mixture of a PCR generally contains buffer to maintain the activity of Taq DNA polymerase, and $MgCl_2$ to enhance the activity of enzyme, mixture of dNTPs, and some other additives to increase the inhibitor tolerance capacity of the *in vitro* DNA replication process. Upon receipt master mix should be stored at −30 to −10 °C. However, after its first use it can be kept at 2–10 °C for its stable storage for 6 months.

3.2 PowerPlex® Fusion 6C 5× Primer Pair Mix

Primer designing is a crucial step of any PCR setup. However, as we use the PCR kit, the primers are designed and optimized for the said reaction by the manufacturer. For multiplexing, a primer set for all the loci to be amplified is included in the mixture. As the next step is detection of the amplified fragments by capillary electrophoresis all the primer sets are labeled with various dye sets. The PowerPlex® Fusion 6C kit amplifies 27 genetic loci at the same time. Hence, 27 pair of primers are present in the primer mix of the kit. Thus, upon receipt primer pair mix should be stored at −30 to −10 °C. However, after its first use it can be kept at 2–10 °C for its stable storage for 6 months. As it is light sensitive, it is recommended to be stored in dark conditions.

3.3 Molecular Biology-Grade Water

Molecular biology-grade water is the form of ultrapure high-quality water which is free from DNase, RNase, and protease contamination. Additionally, no toxic agents such as DEPC should be present in this water. Milli-Q water can be produced in the laboratory or it can be procured from any molecular biology-grade chemical manufacturer.

3.4 Template DNA

Extracted DNA from reference as well as forensic samples can be used as template DNA for amplification of the genetic markers. After estimation of quantity, the DNA should be suitably diluted or concentrated for optimal input in the PCR process, which is ~1 ng. After dilution, the template DNA should be stored at −20 °C till further use. It is always recommended to prepare fresh dilutions of the template DNA in TE buffer to prevent rapid degradation of DNA.

3.5 PCR Machine

A PCR machine provides the physical conditions for *in vitro* replication of DNA in a tube containing all the requisite chemical agents. In the first step of PCR process, denaturation of DNA occurs by raising the temperature to ~95 °C. In the next step the temperature is lowered to as low as ~50 °C which allows the binding of primers at the targeted primer-binding sequence. Further the temperature is increased to ~72 °C for the action of DNA polymerase. These sequential events are repeated for a number of cycles to increase the copy numbers of the targeted DNA fragments. There are many PCR machines available in the market; however, the PCR conditions should be validated internally under the laboratory condition prior to its application in forensic DNA analysis.

4 Procedure

Preparation of Amplification Mixture

1. Calculate the amount of each component of the PCR process using the following table as per your requirement/number of samples (Table 19.1).
2. Thaw PowerPlex® Fusion 6C Master Mix and PowerPlex® Fusion 6C Primer Pair Mix in ice, vortex for 3–5 s, and centrifuge briefly before using.
3. Pipette out the calculated amount of water, master mix, and primer mix in a 1.5 ml microcentrifuge tube marked as Amplification Mixture (AM). It is always recommended to add water first to the tube, followed by PowerPlex® Fusion 6C Master Mix and PowerPlex® Fusion 6C Primer Pair Mix.
4. As soon as the preparation of AM is over, immediately store the tubes containing master mix and primer pair mix in the recommended storage condition.
5. Vortex the amplification mixture for 5–10 s, and then dispense $(5 + 5 + x)$ μl of amplification mixture to each 200 μl PCR tube pre-marked for each sample.
6. Add 1.0 ng of template DNA to each designated PCR tube containing the amplification mixture.
7. In one tube add control DNA in the amplification mix and in another tube add molecular biology-grade water or TE buffer in place of template DNA which will be considered as positive and negative control, respectively.
8. Close the tubes properly and briefly centrifuge the tubes so that all its contents will be present at the bottom of the tube.

Table 19.1
Recommended amounts of reagents required to prepare amplification mixture for PCR

PCR components	Volume per reaction (μl)[a]	No. of reactions	Total volume
PowerPlex® Fusion 6C Master Mix	5.0	A	5A
PowerPlex® Fusion 6C Primer Pair Mix	5.0	A	5A
Molecular biology-grade water	Up to a final volume of 25 μl including template DNA (x)	A	xA
Total volume			$(5 + 5 + x)A$

[a] *Note*: Half volume of the recommended PCR components also gives rise to a good result; however, it should be standardized under your laboratory conditions

Thermal Cycling

9. Place the tubes in the thermal cycler.
10. Set the ramp rate of the thermal cycler to maximum or 100%.
11. Set the thermal cycling protocol as 96 °C for 1 min, followed by 29 cycles of 96 °C for 5 s and 60 °C for 1 min, followed by 60 °C for 10 min, and hold at 4 °C.
12. Wait till completion of the PCR process, which will take around 1–1.25 h.
13. Store the amplified products at −20 °C in dark conditions.

5 Observation and Result

The amplification of the samples can be confirmed after performing capillary electrophoresis and data analysis which has been described in Module 6.

6 Precautions

- Wear gloves while dealing with the reagents and plasticware.
- Do not store the reagents under a light source.
- Always use filtered microtips or aerosol-resistant microtips to prevent any possible contaminations.
- Always put positive and negative control along with the test samples for PCR process.
- As some of the chemicals are hazardous, they should be handled carefully.
- Perform all the works related to PCR setup in a dedicated room using an exclusive set of instruments to avoid contamination.
- Always prepare amplification mixture for extra 10% reactions to avoid any chance of possible pipetting error.
- Make sure not to exceed the volume of template to 20% of the total volume of PCR setup.
- Always ensure to provide the recommended amount of DNA template to the PCR setup. In this purpose, it is useful to ensure the concentration of template DNA by qRT-PCR.
- Do not store the amplification mix for longer time; it may increase the chance of artifacts. Hence, after preparation of amplification mix add template DNA at the earliest followed by immediate thermal cycling.
- Always optimize the reaction setup and thermal cycling conditions as per your laboratory conditions prior to performing any experiment.

7 Troubleshooting

As the results can be evaluated after performing capillary electrophoresis, for troubleshooting of amplification you can refer to the capillary electrophoresis chapter described in Module 6.

Chapter 20

Amplification of Y Chromosome STR Markers by Multiplex PCR

Objective: To amplify the Y chromosome STR markers using multiplex PCR technique.

1 Introduction

Y chromosome and its inheritance from father play a major role in the sex determination of a fetus. Its specific occurrence in males, selective inheritance from father, and high mutation rate are imperative to its investigative and evidentiary value. In male and female mixed samples, Y STR loci typing plays an important role. Some Y STR markers which are closely linked and do not undergo recombination are also useful in paternity determination and establishment of patrilineal relationship. Thus, Y-STR analysis plays an important role along with autosomal STR analysis and a combination of their analyses helps in drawing conclusion with greater confidence.

Amelogenin marker is widely used as a sex-determining marker during forensic analysis. However, the prevalence of deletion at AMELY region of the Y chromosome throughout the globe has raised question on the sole use of amelogenin marker for molecular method of sex determination. Thus, other Y chromosome STR markers are recommended to be used in addition to AMELY for molecular basis of sex determination in humans. Though autosomal STR analysis gives proper identification of an individual, it may not yield a good outcome when mixed-source DNA samples are used. Mostly in sexual assault cases, the samples of victim's origin yield an autosomal profile with excess DNA from the female source in comparison to the male DNA. This is due to the presence of higher number of victim's epithelial cells in the sample than that of sperm cells. Due to preferential amplification by PCR, it becomes difficult to identify the male culprit due to possible allele share

H. R. Dash et al., *Principles and Practices of DNA Analysis: A Laboratory Manual for Forensic DNA Typing*, Springer Protocols Handbooks, https://doi.org/10.1007/978-1-0716-0274-4_20,

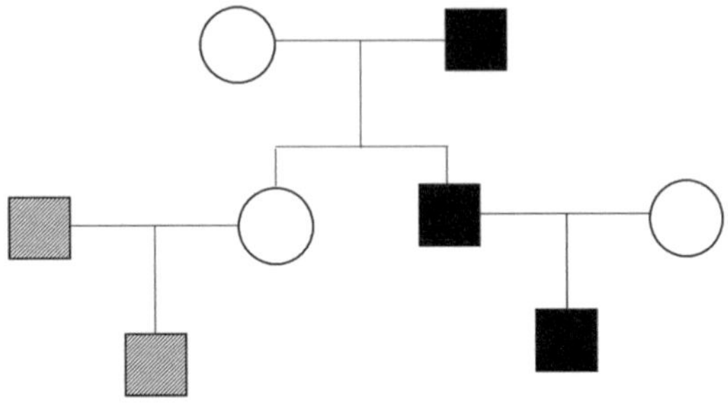

Fig. 20.1 Patrilineal transfer of Y chromosome

between the two DNA contributors. In such cases, Y chromosome STR analysis plays an important role by selective amplification of Y STRs, which is present only in the male culprit's DNA and not present in the female's DNA.

Additionally, as Y chromosome is shared among the paternally related male individuals (Fig. 20.1), Y STR analysis is helpful in solving paternity disputes of male offspring, paternal kinship analysis, as well as familial searching. Y chromosome STR analysis is also useful in drawing inference on geographical origin of an individual as well as in historical and genealogical research. However, the rapidly mutating Y STR markers are not useful in paternity and kinship testing due to their high mutation rates in a generation. In an inbreed population, distinguishing the male relatives is not possible by commercially available Y STR kits. In such cases, for individualization some rapidly mutating Y STR markers with high mutation rates have been discovered. Analysis of a complete set of 13 RM YSTR markers (DYS570, DYS576, DYS627, DYS518, DYS449, DYF387S1ab, DYF399S1, DYF403S1a+b, DYF404S1, DYS526a+b, DYS547, DYS612, and DYS626) is useful in individualizing a suspect which has left his traces in a crime scene than that of his close or distant patrilineal relatives. Though many Y chromosome STR kits are available in the market (list of which has been given in Appendix B), amplification of Y chromosome STR markers has been described in this chapter using Yfiler™ Plus System.

2 Principle

Y chromosome, sized about 60 Mb, is one of the smallest among human chromosomes. About 95% sequences (24 Mb of euchromatin region and 30 Mb of heterochromatin region) of this chromosome are considered to be the non-recombining region (NRY) or male-specific region (MSY). It shows purely paternal inheritance (~99.99%) and remains unaffected by genetic exchanges from X chromosome. Y chromosome also contains the pseudo-autosomal regions (PARs) which are located at the distal end of both p and q

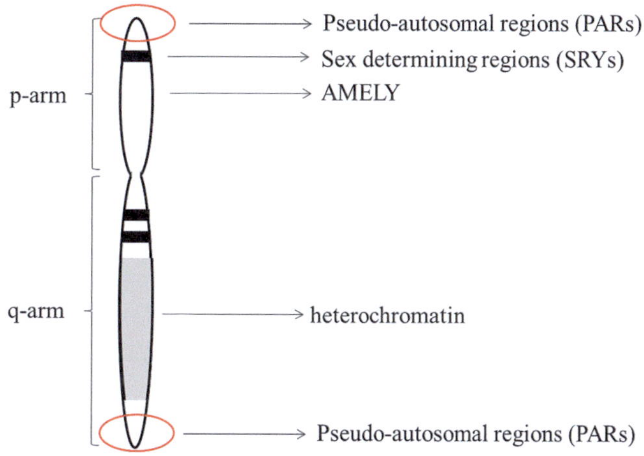

Fig. 20.2 A generalized structure of Y chromosome and its usefulness in forensic DNA analysis

arms. These PARs undergo crossing over and segregation like autosomal loci in one generation to another. It also contains some sex-determining regions (SRY) at band Yp11.3 and amelogenin region which are useful in molecular basis of sex determination in humans (Fig. 20.2).

Initially, a total of 15 Y-STR markers were developed out of which 7 core loci were used for forensic DNA analysis purpose. These sets are being called as the "minimal haplotype Y-STR loci." These minimal haplotype Y-STR loci can distinguish 91–97% of unrelated males in European population. Gradually, the number of STR loci was increased from 7–9 to 17 panels which includes the 7–9 core loci. Most of the STR loci used in this panel are tetranuclear repeats to minimize the stutter products. Gradually various Y-STR kits were developed and validated for forensic use. In this regard, Yfiler™ Plus System amplifies a total of 26 loci which include 7 core loci, i.e., DYS576, DYS389I, DYS635, DYS389II, DYS627, DYS460, DYS458, DYS19, YGATAH4, DYS448, DYS391, DYS456, DYS390, DYS438, DYS392, DYS518, DYS570, DYS437, DYS385 a/b, DYS449, DYS393, DYS439, DYS481, DYF387S1, and DYS533. The included markers also contain three highly polymorphic STR loci, i.e., DYS460, DYS481, and DYS533 and seven rapidly mutating loci, i.e., DYF387S1a/b, DYS449, DYS518, DYS570, DYS576, and DYS627, with >1% of mutation rates to allow differentiation between related males.

A cycle of 30 PCR conditions has been optimized for the use of Yfiler™ Plus kit when a mixture of 1 ng male DNA and 1 μg female DNA is used. The kit has been validated to generate a complete male profile when as low as 250 pg of male DNA is present in a background of 3 μg of female DNA. Yfiler™ Plus works fine when the male-to-female DNA ratio is 1:1000. The kit does not show any cross-reactivity with various non-primates studied, i.e., bovine, chicken, dog, horse, mouse, pig, rabbit, rat, and sheep. A study on the role of potential PCR inhibitors suggests that hematin and humic acids at a concentration of 400 μM and 100 ng/μl do not affect the obtained result. The concordance study also fits well with both the results of Yfiler™ plus and Yfiler kit in four ethnicities, i.e., African American, Caucasian, Hispanic, and Asian. The kit works on any model of Genetic Analyzer that works on six-dye chemistry which includes 3130*xl*, 3500, and 3500xL; the six dyes being used include 6-FAM™ (blue), VIC™ (green), NED™ (yellow), TAZ™ (red), SID™ (purple), and LIZ™ (orange).

3 Reagents/Materials Required and Their Role

3.1 Yfiler™ Plus Master Mix

It contains $MgCl_2$ and mixture of dNTPs, bovine serum albumin (BSA), polymerase enzyme, and 0.05% sodium azide in buffer and salts. Upon the receipt of Yfiler™ Plus Master Mix it should be stored at −25 to −15 °C; however, after first use it can be stored at 2–8 °C.

3.2 Yfiler™ Plus Primer Set

It contains locus-specific 6-FAM™, VIC™, NED™, TAZ™, and SID™ dye-labeled and -unlabeled primers in buffer. During synthesis of primers for DYS389I/II, DYS635, DYS627, DYS19, YGATAH4, DYS448, DYS391, DYS390, DYS438, DYS391, DYS390, DYS438, DYS392, DYS518, DYS437, and DYS449 non-nucleotide linkers are used which are placed between the primers and fluorescent dyes that are used during synthesis of respective oligonucleotide primers. The fundamental behind the use of non-nucleotide linkers is that they provide reproducible positioning of alleles by facilitating inter-locus spacing and simultaneous amplification and efficient separation of the 27 Y-STR loci. Upon the receipt of Yfiler™ Plus Primer Set, it should be stored at −25 to −15 °C; however, after first use it can be stored at 2–8 °C with protection from light.

3.3 Molecular Biology-Grade Water

Molecular biology-grade water is the form of ultrapure high-quality water which is free from DNase, RNase, and protease contamination. Additionally, no toxic agents such as DEPC should be present in this water. Milli-Q water can be produced in the laboratory or it can be procured from any molecular biology-grade chemical manufacturer.

3.4 Template DNA

Extracted DNA from reference as well as forensic samples can be used as template DNA for amplification of the genetic markers. After estimation of quantity, the DNA should be suitably diluted or concentrated for optimal input in the PCR process, which is ~1 ng. After dilution, the template DNA should be stored at −20 °C till further use. It is always recommended to prepare dilutions of the template DNA in TE buffer to prevent rapid degradation of DNA. Additionally, the quality of the template DNA can be maintained by adding 20 μg/ml of glycogen in the TE buffer.

3.5 PCR Machine

A PCR machine provides the physical conditions for *in vitro* replication of DNA in a tube containing all the requisite chemical agents. In the first step of PCR process, denaturation of DNA occurs by raising the temperature to ~95 °C. In the next step the temperature is lowered to as low as ~50 °C which allows the binding of primers at the targeted primer-binding sequence. Further the temperature is increased to ~72 °C for the action of DNA polymerase. These sequential events are repeated for a number of cycles to increase the copy numbers of the targeted DNA fragments. Many PCR machines are available in the market; however, the PCR conditions should be validated internally under the laboratory condition prior to its application in forensic DNA analysis.

4 Procedure

Preparation Amplification Mixture

1. Calculate the amount of each component of the PCR process using the following table as per your requirement/number of samples (Table 20.1).
2. Thaw Yfiler™ Plus Master Mix and Yfiler™ Plus Primer set in ice, vortex for 3–5 s, and centrifuge briefly before using.
3. Pipette out the calculated amount of water, master mix, and primer set in a 1.5 ml microcentrifuge tube marked as Amplification Mixture (AM). It is always recommended to add water first to the tube, followed by Yfiler™ Plus Master Mix and Yfiler™ Plus Primer set.
4. As soon as the preparation of AM is over, immediately store the tubes containing master mix and primer set in the recommended storage condition.
5. Vortex the amplification mixture for 5–10 s, and then dispense $(10 + 5 + x)$ μl of amplification mixture to each 200 μl PCR tube pre-marked for each sample.

Table 20.1
Recommended amounts of reagents required to prepare amplification mixture for PCR

PCR components	Volume per reaction (μl)[a]	No. of reactions	Total volume
Yfiler™ Plus Master Mix	10.0	A	10A
Yfiler™ Plus Primer set	5.0	A	5A
Molecular biology-grade water	Up to a final volume of 25 μl including template DNA (x)	A	xA
Total volume			$(10 + 5 + x)A$

[a] *Note*: Half volume of the recommended PCR components also gives rise to a good result; however, it should be standardized under your laboratory conditions

6. Add 1.0 ng of template DNA to each designated PCR tube containing the amplification mixture.
7. In one tube add control DNA in the amplification mix and in another tube add molecular biology-grade water or TE buffer in place of template DNA which will be considered as positive and negative control, respectively.
8. Close the tubes properly and briefly centrifuge the tubes so that all its contents will be present at the bottom of the tube.

Thermal Cycling

9. Place the tubes in the thermal cycler.
10. Set the ramp rate of the thermal cycler to maximum or 100%.
11. Set the thermal cycling protocol as 95 °C for 1 min, followed by 30 cycles of 94 °C for 4 s and 61.5 °C for 1 min, followed by 60 °C for 22 min, and hold at 4 °C.
12. Wait till completion of the PCR process, which will take around 1–1.25 h.
13. Store the amplified products at −20 °C in dark conditions.

5 Observation and Result

The amplification of the samples can be confirmed after performing capillary electrophoresis and data analysis which has been described in Module 6.

6 Precautions

- Wear gloves while dealing with the reagents and plasticware.
- Do not store the reagents under a light source.
- Always use filtered microtips or aerosol-resistant microtips to prevent any possible contaminations.
- Always put positive and negative control along with the test samples for PCR process.
- As some of the chemicals are hazardous, they should be handled carefully.
- Do all the works relating to PCR setup in a dedicated room using a dedicated set of instruments to avoid contamination.
- Always prepare amplification mixture for extra 10% reactions to avoid any chance of possible pipetting error.
- Make sure not to exceed the volume of template to 20% of the total volume of PCR setup.

- Always ensure to provide the recommended amount of DNA template to the PCR setup. In this purpose, it is useful to ensure the concentration of template DNA by qRT-PCR.
- Do not store the amplification mix for longer time; it may increase the chance of artifacts. Hence, after preparation of amplification mix add template DNA at the earliest followed by immediate thermal cycling.
- Always optimize the reaction setup and thermal cycling conditions as per your laboratory conditions prior to performing any experiments.
- For storage of PCR-amplified products for <2 weeks, place them at 2–8 °C; however for long-term storage (>2 weeks) keep them at −25 °C to −15 °C.

7 Troubleshooting

As the results can be evaluated after performing capillary electrophoresis, for troubleshooting of amplification of Y STR markers you can refer to the capillary electrophoresis chapter described in Module 6.

Chapter 21

Amplification of X Chromosome STR Markers by Multiplex PCR

Objective: To amplify the X chromosome STR markers using multiplex PCR technique.

1 Introduction

STR techniques are widely used in forensic DNA typing for solving various types of cases but not limited to paternity dispute, identification, murder, as well as sexual assault cases. However, in certain complex cases such as kinship analysis as well as cases with limitations of reference samples, autosomal STR analysis alone cannot give a definitive conclusion. In such a scenario, analysis of STRs located on sex chromosomes becomes imperative. Thus, analysis of X-STRs plays a significant role in forensic DNA typing. As men are hemizygous and women are dizygous, men receive single X from their mother. This unique inheritance pattern of X chromosome makes its suitability in deciphering the deficient paternity cases. For a disputed girl child, X-STR analysis between the father and daughter can be performed. However, it cannot be performed for a male child in paternity dispute case due to non-inheritance of X chromosome from father.

The analysis of X-STR in supplement with autosomal STR analysis increases the discrimination power besides affecting the paternity exclusion probability. It is useful in the analysis of complex kinship cases such as paternal half-sisters, paternal aunt/uncle-niece, and maternal uncle-nephew. X-STR analysis is highly useful in incestuous relationships involving grandparent, half-sibling, and uncle/aunt by distinguishing between such relationships. However, the major limitation of X-STR analysis is that as all X-STR markers are present on same X-chromosome, it leads to linkage disequilibrium among them due to linkage. Additionally, the currently available X-STR kits harbor a limited number of markers and generate a large

H. R. Dash et al., *Principles and Practices of DNA Analysis: A Laboratory Manual for Forensic DNA Typing*, Springer Protocols Handbooks, https://doi.org/10.1007/978-1-0716-0274-4_21, © Springer Science+Business Media, LLC, part of Springer Nature 2020

amplicon size. This leads to their inappropriateness in forensic applications. Though few kits are commercially available till now (for detailed list see Appendix B), amplification of X-STRs has been described in this chapter using Investigator Argus X-12 kit.

2 Principle

Human X chromosome is comprised of around 155 million base pairs in length which contains almost 1100 genes. It constitutes 5% and 2.5% of the total DNA of a female and male cell, respectively. In males, the tips of X-chromosome recombine with pseudo-autosomal region of the Y chromosome so that its segregation is maintained during cell division. However, both the X chromosomes in the female recombine just like any other autosomal chromosome during meiosis. Such unique biological nature and pattern of inheritance increase the usefulness of X-chromosome STR typing in various cases, i.e., complex kinship testing involving female samples, motherless paternity dispute cases, half-sister testing, comparison of grandparent and grandchildren, and paternity establishment in incest cases (Fig. 21.1).

In this regard, Argus X-12 kit provides a multiplex system for simultaneous amplification of 12 X-STR markers, namely DXS7132, DXS7423, DXS8378, DXS10074, DXS10079, DXS10101, DXS10103, DXS10134, DXS10135, DXS10146, DXS10148, and HPRTB, besides amelogenin marker for sex determination. The X-STRs present in less than 1 cm distance are considered as a linkage group which are inherited together during recombination. These 12 markers are inherited as four linkage groups (i.e., group 1:

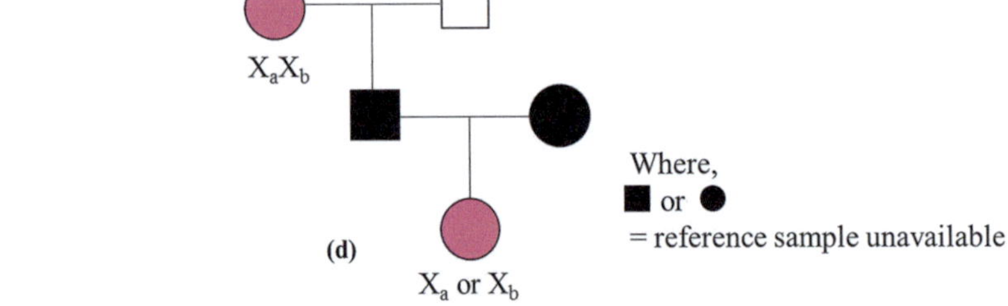

Fig. 21.1 Few types of cases where X-STR analysis is useful: (**a**) disputed female child with mother unavailable, (**b**) disputed male child with father unavailable, (**c**) disputed female child with father unavailable, (**d**) identification of female individual with available reference sample of paternal grandmother

DXS8378, DXS10135, and DXS10148; group 2: DXS7132, DXS10074, and DXS10079; group 3: HPRTB, DXS10101, and DXS10103; group 4: DXS7423, DXS10134, and DXS10146). This reflects that the result of each set of these three markers can be treated as a different haplotype in genotyping. Due to the presence of these linkage groups, the result of X-STR analysis can be used as an alternative tool for the analysis of opposite-sex siblings with increased specificity.

The kit requires a template DNA amount of 0.2–0.5 ng for optimum result. It works on any genotyping platform which can accommodate five-dye chemistry or more as the primers of the markers are labeled with four dyes, i.e., 6-FAM™ (blue), BTG (green), BTY (yellow), and BTR (red). Thus, it can be used on ABI PRISM® 310 and 3100 and Applied Biosystems 3130 and 3500 genetic analyzers. A simulated study on the effect of various inhibitors such as humic acid (200 ng/μl), hematin (500 μM), tannic acid (3000 ng/μl), indigo carmine (6 mM), collagen (150 ng/μl), and calcium (2 mM) suggests the suitability of the kit in the presence of various potential PCR inhibitors. The kit works fine in the degraded samples with as low as 150 bp of fragment-sized DNA. Besides chimpanzees, bonobos, orangutans, and gorillas, no other animals give rise to a reproducible peaks using Argus X-12 kit.

3 Reagents/Materials Required and Their Role

3.1 Reaction Mix A

It contains dNTPs mixture, $MgCl_2$, and bovine serum albumin (BSA). dNTPs serve as the building blocks for *in vitro* DNA synthesis by PCR. $MgCl_2$ acts as a cofactor and enhances the productivity of Taq polymerase. BSA stabilizes the enzyme during storage and provides an environment free from nucleases. Additionally, BSA also prevents adhering of Taq polymerase to the wall of the PCR tubes. Upon the receipt of reaction mix A it should be stored at −25 to −15 °C.

3.2 Primer Mix

A set of primers for each locus are present in this mixture. Each primer set is labeled with a specific fluorescent dye, i.e., 6-FAM™ (amelogenin, DXS10103, DXS8378, DXS7132, DXS10134), BTG (DXS10074, DXS10101, DXS10135), BTY (DXS7423, DXS10146, DXS10079), and BTR (HPRTB, DXS10148). Upon the receipt of the primer mix, it should be stored at −25 to −15 °C with protection from light.

3.3 Multi-Taq2 DNA Polymerase

It contains Taq polymerase which attaches the dNTPs to a DNA template by extending the length of primer. Thus, it helps in a sister strand synthesis. Taq polymerase isolated from a thermophilic bacterium *Thermus aquaticus* is chosen for this purpose as it can withstand a temperature required for the PCR process. It should be stored at −20 °C for its better performance.

3.4 Molecular Biology-Grade Water

Molecular biology-grade water is the form of ultrapure high-quality water which is free from DNase, RNase, and protease contamination. Additionally, no toxic agents such as DEPC should be present in this water. Milli-Q water can be produced in the laboratory or it can be procured from any molecular biology-grade chemical manufacturer.

3.5 Template DNA

Extracted DNA from reference as well as forensic samples can be used as template DNA for amplification of the genetic markers. After estimation of quantity, the DNA should be suitably diluted or concentrated for optimal input in the PCR process, which is ~1 ng. After dilution, the template DNA should be stored at −20 °C till further use. It is always recommended to prepare dilutions of the template DNA in TE buffer to prevent rapid degradation.

3.6 PCR Machine

A PCR machine provides the physical conditions for *in vitro* replication of DNA in a tube containing all the requisite chemical agents. In the first step of PCR process, denaturation of DNA occurs by raising the temperature to ~95 °C. In the next step the temperature is lowered to as low as ~50 °C which allows the binding of primers at the targeted primer-binding sequence. Further the temperature is increased to ~72 °C for the action of DNA polymerase. These sequential events are repeated for a number of cycles to increase the copy numbers of the targeted DNA fragments. There are many PCR machines available in the market; however, the PCR conditions should be validated internally under the laboratory condition prior to its application in forensic DNA analysis.

4 Procedure

Preparation Amplification Mixture

1. Calculate the amount of each component of the PCR process using the following table as per your requirement/number of samples (Table 21.1).
2. Thaw reaction mix A and primer mix in ice, vortex for 3–5 s, and centrifuge briefly before using.
3. Pipette out the calculated amount of molecular biology-grade water, reaction mix A, primer mix, and multi-Taq2 DNA polymerase in a 1.5 ml microcentrifuge tube marked as Amplification Mixture (AM). It is always recommended to add water first to the tube, followed by reaction mix A, primer mix, and multi-Taq2 DNA polymerase.
4. As soon as the preparation of AM is over, immediately store the tubes containing reaction mix A, primer mix, and multi-Taq2 DNA polymerase in the recommended storage condition.

Table 21.1
Recommended amounts of reagents required to prepare amplification mixture for PCR

PCR components	Volume per reaction (μl)	No. of reactions	Total volume
Reaction Mix A	5.0	A	5A
Primer Mix	2.5	A	2.5A
Multi-Taq2 DNA polymerase	0.6	A	0.6A
Molecular biology-grade water	Up to a final volume of 25 μl including template DNA (x)	A	xA
Total volume			*(5 + 2.5 + 0.6 + x)A*

5. Vortex the amplification mixture for 5–10 s, and then dispense (5 + 2.5 + 0.6 + x) μl of amplification mixture to each 200 μl PCR tube pre-marked for each sample.
6. Add 1.0 ng of template DNA to each designated PCR tube containing the amplification mixture.
7. In one tube add control DNA in the amplification mix and in another tube add molecular biology-grade water or TE buffer in place of template DNA which will be considered as positive and negative control, respectively.
8. Close the tubes properly and briefly centrifuge the tubes so that all its contents will be present at the bottom of the tube.

Thermal Cycling

9. Place the tubes in the thermal cycler.
10. Set the ramp rate of the thermal cycler to maximum or 100%.
11. Set the thermal cycling protocol as 94 °C for 4 min; followed by 5 cycles of 96 °C for 30 s, 63 °C for 120 s, and 72 °C for 75 s; followed by 94 °C for 30 s, 60 °C for 120 s, and 72 °C for 75 s; and hold at 4 °C.
12. Wait till completion of the PCR process.
13. Store the amplified products at −20 °C in dark conditions or immediately process for electrophoresis.

5 Observation and Result

The amplification of the samples can be confirmed after performing capillary electrophoresis and data analysis which has been described in Module 6.

6 Precautions

- Wear gloves while dealing with the reagents and plasticware.
- Do not store the reagents under a light source.
- Always use filtered microtips or aerosol-resistant microtips to prevent any possible contaminations.
- Always put positive and negative control along with the test samples for PCR process.
- As some of the chemicals are hazardous, they should be handled carefully.
- Do all the works relating to PCR setup in a dedicated room using a dedicated set of instruments to avoid contamination.
- Always prepare amplification mixture for extra 10% reactions to avoid any chance of possible pipetting error.
- Make sure not to exceed the volume of template to 20% of the total volume of PCR setup.
- Always ensure to provide the recommended amount of DNA template to the PCR setup. For this purpose, it is useful to ensure the concentration of template DNA by qRT-PCR.
- Do not store the amplification mix for longer time; it may increase the chance of artifacts. Hence, after preparation of amplification mix add template DNA at the earliest followed by immediate thermal cycling.
- Always optimize the reaction setup and thermal cycling conditions as per your laboratory conditions prior to performing any experiments.
- For storage of PCR-amplified products for <2 weeks, place them at 2–8 °C; however for long-term storage (>2 weeks) keep them at −25 to −15 °C.

7 Troubleshooting

As the results can be evaluated after performing capillary electrophoresis, for troubleshooting of amplification you can refer to the capillary electrophoresis chapter described in Module 6.

Chapter 22

Amplification of Autosomal Mini-STR Markers from Degraded Samples

Objective: To amplify the autosomal mini-STR markers by multiplex PCR technique.

1 Introduction

Most of the biological samples received for DNA examination are degraded in nature. Due to exposure to varied environmental conditions, the samples also contain many potential PCR inhibitors or environmental insults. The most common damage to DNA in forensic and ancient analysis relies on the degradation of DNA by both endogenous and exogenous nuclease activity as well as nonenzymatic hydrolytic cleavage which generates strand breaks. In such compromised samples, a partial DNA profile is highly expected as amplification of large-sized STR loci is carried out during PCR process using routine STR kits. Such partial DNA profiles do not value much from forensic point of view as they do not provide sufficient power of discrimination to either include or exclude a suspect. In such cases, an advanced technique can be exploited for amplification of STR markers using specially designed primers. These primers preferentially target the larger STR loci in such a way to generate a smaller amplification product.

Mini-STR primers possess the capability to magnify the STR loci which increases the chance of successive amplification from the degraded samples. Thus, the mini-STR technology can generate a DNA profile from the compromised samples with increased sensitivity. In this way, analysis of mini-STRs is highly useful in providing genetic data from degraded samples with extremely low-quantity as well as -quality DNA. Additionally, the quality of DNA profiling result is increased manyfold by using the mini-STR technology. Mini-STR analysis is capable of extracting genetic information

H. R. Dash et al., *Principles and Practices of DNA Analysis: A Laboratory Manual for Forensic DNA Typing*, Springer Protocols Handbooks, https://doi.org/10.1007/978-1-0716-0274-4_22,

from STR loci with allele size <150 bp more efficiently. The adverse quality of STR profiles witnessed by high peak height imbalance in degraded samples can be minimized using mini-STR technology. This technology may open a new era in forensic DNA typing as previously unsolvable cases due to nonavailability of suitable DNA can be resolved. The protocol for amplifying mini-STRs from degraded samples using AmpF*l*STR™ MiniFiler™ PCR amplification kit has been described in this chapter.

2 Principle

Most of the commonly used STR loci contain a large number of repeats and are unable to cause amplification when less than 150 bp size of a DNA fragment is present. In such degraded DNA samples, there are two ways to produce a result, i.e., (a) application of new polymorphic markers with smaller repeat units and narrow allele range or (b) moving forward the forward and reverse primers to generate a low amplicon size. The fundamental of mini-STRs is to increase the proximity of the forward and reverse primers to the STR region (Fig. 22.1). In this way, the amplicon size of STRs becomes less and genetic information can be recovered from the degraded samples.

In this regard, the AmpF*l*STR™ MiniFiler™ PCR amplification kit is capable of amplifying eight autosomal loci simultaneously, i.e., D13S317, D7S820, D2S1338, D21S11, D16S539, D18S51, CSF1PO, and FGA, besides the sex-determining marker amelogenin. This kit utilizes the non-nucleotide linkers to reduce the amplicon size of large loci which are more prone to allele dropout. These linkers are capable of positioning the alleles which facilitates inter-locus spacing. Additionally, combinatorial use of five-dye fluorescent system (6-FAM™, VIC™, NED™, PET™, and LIZ™) and linkers allows simultaneous amplification and separation of these STR markers. In comparison to the other commercially available kits, the markers show average 58% and 48% size reduction in smallest and largest allele, respectively, except D21S11. Amplicon length of the markers has been reduced significantly to –129 bp (D7S820), –99 bp (D13S317), –33 bp (D21S11), –183 bp (D2S1338), –168 bp (D18S51), –157 bp (D16S539), –87 bp (FGA), and –201 bp (CSF1PO). In this way, the kit is highly useful in tough samples such as bone, old stains, hairs, and swabs containing low level of DNA.

Many population studies have demonstrated the concordance of the AmpF*l*STR™ MiniFiler™ PCR amplification kit result with other commercially available kits. The validation experiments of the kit as per SWAGDAM guidelines advocate its usefulness in forensic DNA typing. Though full profile has been obtained using 0.125 ng of template DNA, the optimal DNA quantity for AmpF*l*STR™

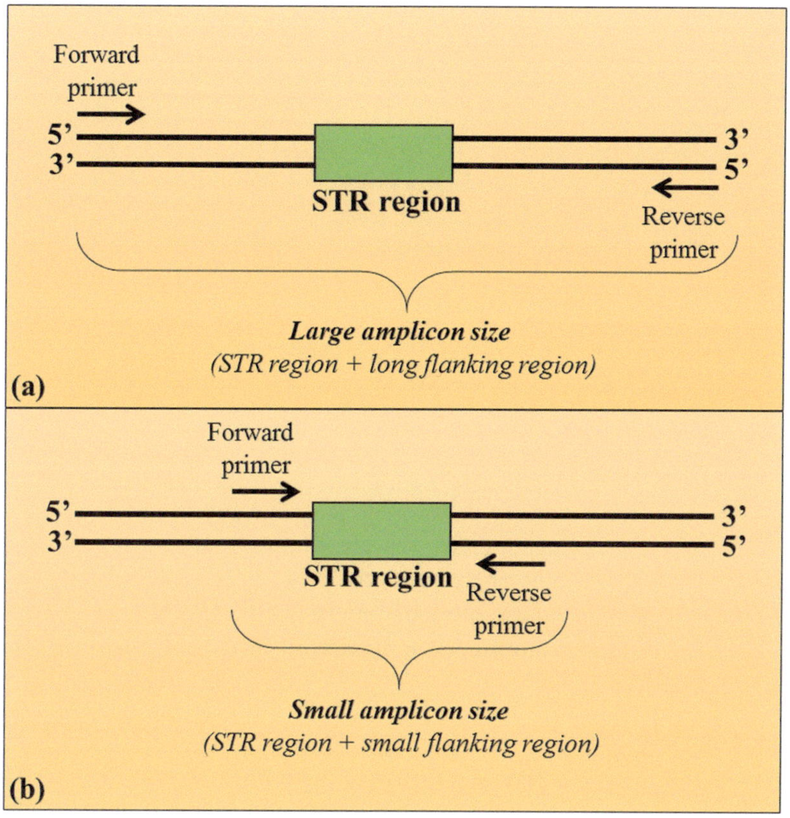

Fig. 22.1 Selection of primers for (**a**) large amplicon size in conventional STR analysis and (**b**) small amplicon size in mini-STR analysis

MiniFiler™ PCR amplification kit has been recommended to be 0.5–0.75 ng. A full profile has been generated when a good-quality DNA is digested with 5 U of *DNaseI* for 20 min, suggesting its usefulness in highly degraded sample.

3 Reagents/Materials Required and Their Role

3.1 AmpFlSTR™ MiniFiler™ Master Mix

It contains polymerase enzyme, $MgCl_2$, dNTPs, carrier proteins such as BSA, and sodium azide. dNTPs serve as the building blocks for *in vitro* DNA synthesis by PCR. $MgCl_2$ acts as a cofactor and enhances the productivity of Taq polymerase. Taq polymerase attaches the dNTPs to a DNA template by extending the length of the primer. Carrier protein such as BSA stabilizes the enzyme during storage and provides an environment free from nucleases. Additionally, BSA also prevents adhering of Taq polymerase to the wall of the PCR tubes. Sodium azide is a preservative used in buffer to prevent exogenous contamination. Upon the receipt of reaction mix A it should be stored at −25 to −15 °C and after initial use it should be stored at 2–8 °C.

3.2 AmpF*l*STR™ MiniFiler™ Primer Set

A set of primers for each locus are present in this mixture. This contains the forward and reverse primers to amplify the specific human targets categorically. Each primer set is labeled with a specific fluorescent dye, i.e., 6-FAM™ (D13S317, D7S820), VIC™ (amelogenin, D2S1338, D21S11), NED™ (D16S539, D18S51), PET™ (CSF1PO, FGA), and LIZ™ (Size Standard). Upon the receipt of the primer mix, it should be stored at −25 to −15 °C with protection from light.

3.3 Molecular Biology-Grade Water

Molecular biology-grade water is the form of ultrapure high-quality water which is free from DNase, RNase, and protease contamination. Additionally, no toxic agents such as DEPC should be present in this water. Milli-Q water can be produced in the laboratory or it can be procured from any molecular biology-grade chemical manufacturer.

3.4 Template DNA

Extracted DNA from reference as well as forensic samples can be used as template DNA for amplification of the genetic markers. After estimation of quantity, the DNA should be suitably diluted or concentrated for optimal input in the PCR process, which is 0.5–0.75 ng. After dilution, the template DNA should be stored at −20 °C till further use. It is always recommended to prepare dilutions of the template DNA in TE buffer to prevent rapid degradation.

3.5 PCR Machine

A PCR machine provides the physical conditions for *in vitro* replication of DNA in a tube containing all the requisite chemical agents. In the first step of PCR process, denaturation of DNA occurs by raising the temperature to ~95 °C. In the next step the temperature is lowered to as low as ~50 °C which allows the binding of primers at the targeted primer-binding sequence. Further the temperature is increased to ~72 °C for the action of DNA polymerase. These sequential events are repeated for a number of cycles to increase the copy numbers of the targeted DNA fragments. There are many PCR machines available in the market; however, the PCR conditions should be validated internally under the laboratory condition prior to its application in forensic DNA analysis.

4 Procedure

Preparation Amplification Mixture

1. Calculate the amount of each component of the PCR process using the following table as per your requirement/number of samples (Table 22.1).
2. Thaw AmpF*l*STR™ MiniFiler™ Master Mix and AmpF*l*STR™ MiniFiler™ Primer Set in ice, vortex for 3–5 s, and centrifuge briefly before using.

Table 22.1
Recommended amounts of reagents required to prepare amplification mixture for PCR

PCR components	Volume per reaction (μl)	No. of reactions	Total volume
AmpF/STR™ MiniFiler™ Master Mix	10.0	A	10A
AmpF/STR™ MiniFiler™ Primer Set	5.0	A	5A
Molecular biology-grade water	Up to a final volume of 25 μl including template DNA (x)	A	xA
Total volume			*(10 + 5 + x)A*

3. Pipette out the calculated amount of molecular biology-grade water, AmpF/STR™ MiniFiler™ Master Mix, and AmpF/STR™ MiniFiler™ Primer Set in a 1.5 ml microcentrifuge tube marked as Amplification Mixture (AM). It is always recommended to add water first to the tube, followed by AmpF/STR™ MiniFiler™ Master Mix and AmpF/STR™ MiniFiler™ Primer Set.
4. As soon as the preparation of AM is over, immediately store the tubes containing AmpF/STR™ MiniFiler™ Master Mix and AmpF/STR™ MiniFiler™ Primer Set in the recommended storage condition.
5. Vortex the amplification mixture for 5–10 s, and then dispense (10 + 5 + x) μl of amplification mixture to each 200 μl PCR tube pre-marked for each sample.
6. Add 0.5–0.75 ng of template DNA to each designated PCR tube containing the amplification mixture.
7. In one tube add control DNA in the amplification mix and in another tube add molecular biology-grade water or TE buffer in place of template DNA which will be considered as positive and negative control, respectively.
8. Close the tubes properly and briefly centrifuge the tubes so that all its contents will be present at the bottom of the tube.

Thermal Cycling

9. Place the tubes in the thermal cycler.
10. Set the ramp rate of the thermal cycler to maximum or 100%.
11. Set the thermal cycling protocol as 95 °C for 11 min; followed by 30 cycles of 94 °C for 20 s, 59 °C for 2 min, and 72 °C for 1 min; followed by 60 °C for 45 min; and hold at 4 °C.
12. Wait till completion of the PCR process.
13. Store the amplified products at −20 °C in dark conditions or immediately process for electrophoresis.

5 Observation and Result

The amplification of the samples can be confirmed after performing capillary electrophoresis and data analysis which has been described in Module 6.

6 Precautions

- Wear gloves while dealing with the reagents and plasticware.
- Do not store the reagents under a light source.
- Always use filtered microtips or aerosol-resistant microtips to prevent any possible contaminations.
- Always put positive and negative control along with the test samples for PCR process.
- As some of the chemicals are hazardous, they should be handled carefully.
- Do all the works relating to PCR setup in a dedicated room using a dedicated set of instruments to avoid contamination.
- Always prepare amplification mixture for extra 10% reactions to avoid any chance of possible pipetting error.
- Make sure not to exceed the volume of template to 20% of the total volume of PCR setup.
- Always ensure to provide the recommended amount of DNA template to the PCR setup. For this purpose, it is useful to ensure the concentration of template DNA by qRT-PCR.
- Do not store the amplification mix for longer time; it may increase the chance of artifacts. Hence, after preparation of amplification mix add template DNA at the earliest followed by immediate thermal cycling.
- Always optimize the reaction setup and thermal cycling conditions as per your laboratory conditions prior to performing any experiments.
- For storage of PCR-amplified products for <2 weeks, place them at 2–8 °C; however for long-term storage (>2 weeks) keep them at −25 to −15 °C.

7 Troubleshooting

As the results can be evaluated after performing capillary electrophoresis, for troubleshooting of amplification you can refer to the capillary electrophoresis chapter described in Module 6.

Part VI

Genotyping and Analysis of Results

Chapter 23

Separation of Amplified DNA Fragments by Capillary Electrophoresis Using Genetic Analyzer 3500xL

Objective: To separate the amplified fragments by capillary electrophoresis using Genetic Analyzer 3500xL.

1 Introduction

After amplification of genetic loci (STRs) using multiplex process, the amplified fragments need to be analyzed by electrophoresis. In this process, the amplified products get separated on the basis of their fragment sizes and the corresponding alleles can be designated. As the amplified products are multiple in numbers labeled with different fluorescent dyes, a sophisticated, automated instrument is required to separate as well as identify the amplified fragment with great efficiency. In this regard, since the inception of STR-based forensic DNA typing, capillary electrophoresis has become an integral part of the analysis process. It is an advanced version of the electrophoresis process, where the gel slab has been replaced by the narrow capillaries for the efficient separation of nucleic acids. The narrow capillary is capable of performing a versatile analysis at high electric fields, thus decreasing the run time as well as minimizing the overheating problems during the application of high voltage. The technique has become immensely popular among the users due to its capabilities of automation leading to elimination of preparation of gel slabs as well as loading the DNA fragments onto a gel slab. Additionally, consumption of a scanty amount of DNA and software-based detection and analysis of DNA fragments have popularized the capillary electrophoresis process.

Coupling of capillary electrophoresis with automated injection of samples has increased the versatility of the instrument. Invention of instruments with multiple fluorescence detection capabilities has increased the efficiency of the instrument in analyzing multiple

H. R. Dash et al., *Principles and Practices of DNA Analysis: A Laboratory Manual for Forensic DNA Typing*, Springer Protocols Handbooks, https://doi.org/10.1007/978-1-0716-0274-4_23,

genetic fragments in a single run. It has also shown its usefulness in other applications such as sequencing of human genome, gunshots, and explosives residue analysis as well as drug analysis. The capillary number of various instruments varies from 1 to 384, where these many number of samples can be analyzed simultaneously in a single run. Many capillary electrophoresis systems are available in the market (for details see Appendix B); however, ABI Genetic Analyzer 3500xL instrument, a 24-capillary 6-dye chemistry instrument, has been described in this chapter.

2 Principle

A conventional capillary electrophoresis system involves a high-voltage power supply, a system for sample injection, a narrow-width capillary tube, the detector, and the output system (Fig. 23.1). In most of the instruments, a temperature control device is also present to maintain the temperature. Maintenance of temperature is required as an increase in temperature leads to improper electrophoretic mobility and decrease in viscosity of the solution. For application of high voltage, two electrodes are included which initiates the migration of samples from cathode to anode under the influence of electric current inside the capillary. Mostly the capillary is made up of silica which is dipped in cathode and anode buffer in both the sides. Before electrophoresis, the capillary is filled with the polymer which acts as the matrix for the transfer of DNA fragments. A laser beam along with a detector is fitted at the anode end of the capillary to detect the fluorescence of the amplified fragments. Additionally a photomultiplier is connected to gather additional information of the separated fragments.

Samples (amplified DNA) are injected into the capillary array by a technique called as electrokinetic injection. In this process when an electric field is applied to the liquid sample, the negatively charged DNA molecules get dragged by the positive voltage into the capillary. Such a type of injection produces narrow injection zones to allow separation of DNA with high resolution. However, impurities

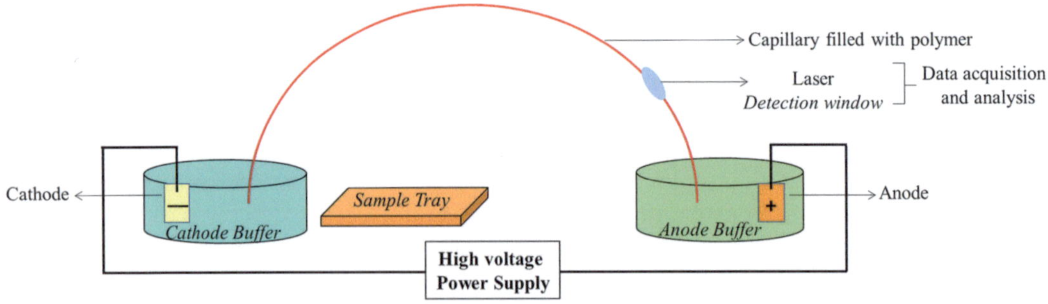

Fig. 23.1 Schematic diagram of capillary electrophoresis instrument along with its working components

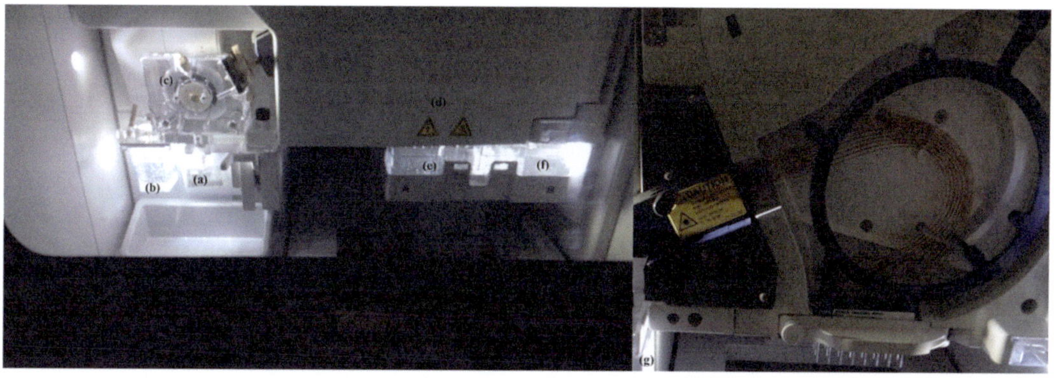

Fig. 23.2 A view of Genetic Analyzer 3500xL: (**a**) Polymer pack, (**b**) outlet buffer reservoir, (**c**) mechanical pump, (**d**) oven door, (**e**) inlet buffer reservoir, (**f**) autosampler, (**g**) inside view of the oven chamber showing 24 capillaries

like chloride ions in DNA solution may compete with the negatively charged DNA molecules to be injected into the capillary by the electrokinetic injection process. In this regard, the quantity of DNA injected into a capillary is dependent on voltage (E), injection time (t), DNA concentration (DNA_{sample}), capillary opening area (πr^2), ionic strength of sample (λ_{sample}) and buffer (λ_{buffer}), electrophoretic mobility (μ_{ep}), and electroosmotic flow (μ_{eof}), as follows:

$$\left[\mathrm{DNA_{injected}}\right] = E \times t\left(\pi r^2\right)\left(\mu_{ep} + \mu_{eof}\right) \times \left[\mathrm{DNA_{sample}}\right] \left(\lambda_{buffer}\right)/\lambda_{sample}$$

After completion of electrophoresis, the fluorescently labeled PCR products are detected by a fluorescent detector either with a photomultiplier tube (PMT) or charge-coupled device (CCD). Thus, the amplified STR alleles are marked as peaks on an electropherogram with the light intensity measured as arbitrary unit, i.e., relative fluorescence units (RFUs). Genetic Analyzer 3500xL is a 24-capillary instrument with reduced power requirements (110 V) and capability of 6-dye detection (Fig. 23.2). Additionally, the monitoring of consumables is carried out by the radio-frequency identification (RFID) tags for ready-to-use and easy maintenance of reagents and consumables.

3 Reagents/Materials Required and Their Role

3.1 PCR-Amplified Product

The products of PCR after amplification of autosomal STRs, Y-STRs, X-STRs, or mini-STRs are required for run in Genetic Analyzer. The PCR products can be immediately processed for electrophoresis or stored at $-20°$ till electrophoresis. It is always recommended to store the PCR products under dark conditions.

3.2 Hi-Di™ Formamide

This is a form of highly deionized formamide used for resuspension of PCR products for electrokinetic injection during capillary electrophoresis. Additionally, it acts as a denaturant and provides stability to the sample during heat denaturation step as well as during the process of capillary electrophoresis. The commonly used POP-4 for fragment analysis contains linear, uncross-linked dimethyl polyacrylamide at a concentration of 4%. Presence of urea in this solution provides added advantage during capillary electrophoresis by creating an environment for keeping the DNA molecules in denatured condition. Before using a polymer pouch, check for the presence of crystals (if any) which may hamper the electrophoresis process. It can be stored at 2–8 °C for up to 1 week; however, for long-term storage up to 3 months, it should be stored at 15 to −25 °C. The best way to use formamide is to prepare small aliquots in 1.5 ml or smaller tubes for routine use and avoiding repeated freeze-thaw or exposure to air.

3.3 Internal Size Standard

Internal size standard is mixed with each sample for processing of samples as well as separation and detection of STR alleles. It is useful in the correct detection of size of the amplified fragments. An additional dye set, mostly the LIZ™, is used for labeling of the size standard. As different samples are mixed with a set of DNA fragments of known size, it can correlate the results from run to run as well as it helps in troubleshooting of the process at the step of either PCR or capillary electrophoresis. Peak height normalization can also be performed using internal size standard which leads to peak height uniformity among capillaries, run, and instruments. It is kit specific. Hence, size standard supplied with the STR amplification kit should be used for electrophoresis of those samples.

3.4 Allelic Ladder

It is an artificial mixture of nucleotide fragments that are of same size to the common alleles available in human population for all the markers analyzed. This plays a major role in allele designation in a STR locus after PCR amplification. Allelic ladder acts as a ruler for each locus, as the amplified fragment will match with the position of the allele present on the ladder to designate it as an allele. Commercial manufacturers supply allelic ladders along with their kits. As the allele range of the genetic markers varies, it is always recommended to use the specific allelic ladder of the specific kit for analysis purpose. The allelic ladder should be transported at −15 to −25 °C whereas while in use it should be stored at 2–8 °C with protection from light. As allelic ladder contains the products from PCR, it should not be stored along with the unamplified DNA.

3.5 Capillary

Capillary is the most vital accessory of the instrument. It is commonly made up of fused silica (SiO_2) and their inner wall contains the hydroxyl groups which are negatively charged at pH 5.0. When electrophoresis buffer with pH 8.0 is used, positive ions from buffer line up with SiO^- along their wall to create a double layer. Under

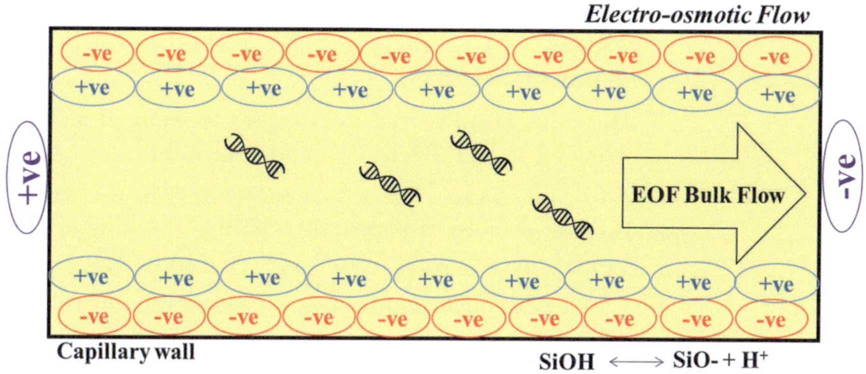

Fig. 23.3 Internal environment of a capillary tube generating the electroosmotic flow

electric current, the positive ions move towards the negative electrode called as electroosmotic flow (EOF) (Fig. 23.3). The EOF may create problem to obtain a DNA separation result with higher reproducibility. Thus, the capillary is chemically modified or dynamically coated to minimize/prevent EOF during electrophoresis. To avoid problems and increase the separation resolution, the capillary should be replaced each time after 100–150 injections.

3.6 Buffers

For electrophoresis in a Genetic Analyzer, anode and cathode buffers are necessary. The common constituents of anode and cathode buffers include 100 mM/l *N*-tris-(hydroxymethyl)-methyl-3-aminopropane-sulfonic acid (TAPS) and 1 mM/l EDTA adjusted to pH 8.0. Additionally to accomplish DNA denaturation and maintain temperature during electrophoresis some additives are also added which include formamide, urea, and 2-pyrrolidinone. After repeated runs the composition of anode and cathode buffers changes due to migration of buffer ions and electrolysis. Thus, it should be changed and replenished with fresh set of buffer within a stipulated time interval.

3.7 Polymer

Commercially available poly-dimethylacrylamide POP-4 and POP-6 polymers are mostly used in Genetic Analyzer 3500xL. It provides a sieving matrix for efficient separation of single-stranded DNA during capillary electrophoresis. Additionally as it suppresses EOF, it increases resolution of amplified fragments. The constituents of POP-4 include mostly di-methylacrylamide (4%), urea (8 mol/l), and 2-pyrrolidinone (5%).

4 Procedure

Plate Preparation

1. Thaw PCR-amplified products, Hi-Di™ Formamide, size standard, and allelic ladder on ice. Vortex size standard and allelic ladder briefly for 10–15 s for proper mixing of the fragments.

2. Pipette out 247 μl of Hi-Di™ Formamide and 13 μl of size standard in a 1.5 ml microcentrifuge tube. Vortex for 10–15 s to mix the reagents.
3. Dispense 10 μl of the mixture in 24 wells of a 96-well plate from 1A to 1H, 2A to 2H, and 3A to 3H.
4. Add 1 μl of the PCR-amplified product in the wells (you can prepare a sheet of paper, designating the sample name to be loaded on each well) from 1A to 1H, 2A to 2H, and 3A to 3G.
5. Add 1 μl of allelic ladder to well 3H of the 96-well plate.
6. Cover the 96-well plate with an appropriate septa.
7. Centrifuge the plate at 5000 rpm for 30 s at room temperature using a plate rotor.
8. Denature the samples by incubating the plate at 95 °C for 3 min followed by immediate freezing on ice for 3 min. Always denature samples just before loading of the plate in the instrument.
9. Place the plate containing samples on the plate base and cover it with plate retainer. Check for the proper alignment of the holes of the plate retainer with the holes of the septa. Click on "Tray" button of the instrument, wait till the indicator turns green, load the plate, and close the door (Fig. 23.4).

Instrument Setup and Preparation of Plate Record

10. Switch on the computer followed by switching on the instrument.
11. Wait till the computer finishes booting and the connection between the computer and instrument is established. Completion of booting can be confirmed by "active" of the status icon of the 3500 server monitor present at the right lower corner of the desktop (Fig. 23.5).

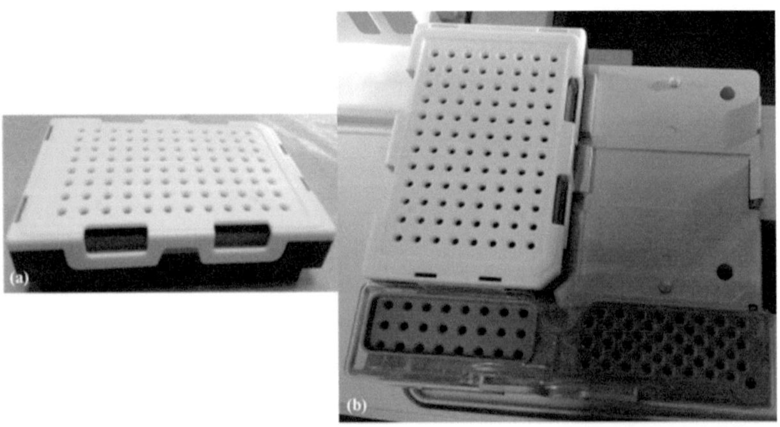

Fig. 23.4 (**a**) A 96-well plate with loaded samples, covered with septa placed on a plate base and covered with plate retainer; (**b**) loading of the plate in the instrument

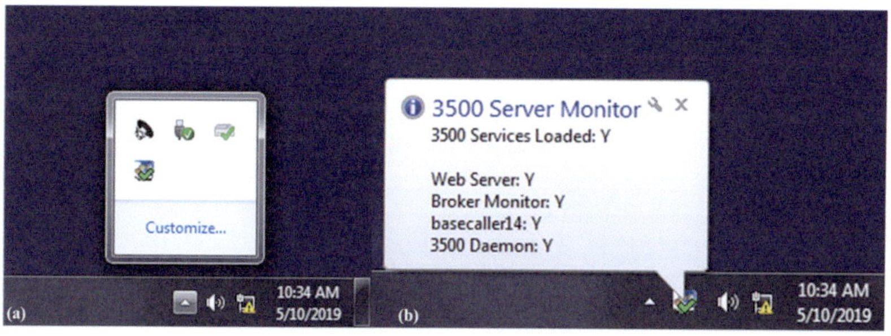

Fig. 23.5 "Active icon" confirming the establishment of connection between the instrument and computer as (**a**) Active icon and (**b**) 3500 server. Monitor activation confirming the establishment of connection between the instrument and computer

12. Launch the 3500 Data collection software by using the designated "user id" and "password."
13. The Dashboard is displayed after opening the software (Fig. 23.6). This Dashboard shows information and status of the instrument, consumables, and maintenance.
14. From the Dashboard pick "Create New Plate"; enter a "plate name"; select the number of wells as "96," plate type as "HID," capillary length as "36 cm," and polymer as "POP-4" (Fig. 23.7).
15. Go to "Assign Plate Contents"; a 96-well format will appear on the screen (Fig. 23.8).
16. Go to "Assays"; click on "Add from library"; select the suitable assay you want to perform, for electrophoresis of amplicons amplified using Global Filer™ kit; select "GF+Norm_POP4-xl"; click on "Add to Plate"; and then close the window (Fig. 23.9).
17. Go to "File Name Convention"; click on "Add from library"; select the suitable file name convention you want to perform; click on "Add to Plate"; and then close the window (Fig. 23.10).
18. Go to "Result Groups"; click on "Add from library"; select the suitable Result Group where you want to keep your results; click on "Add to Plate,"; and then close the window (Fig. 23.11).
19. Click on the corresponding position on the 96-well format that appears on the screen and add information of the samples loaded at that position. Make sure to add information of ladder loaded at well 3H (Fig. 23.12).
20. Select all the wells with information added to each well and click on the box present near the selected Assay, File name convention, and Result Group (Fig. 23.13).
21. Go to customize Sample Info, select all the samples except ladder, and assign "Sample" from the drop-down menu.

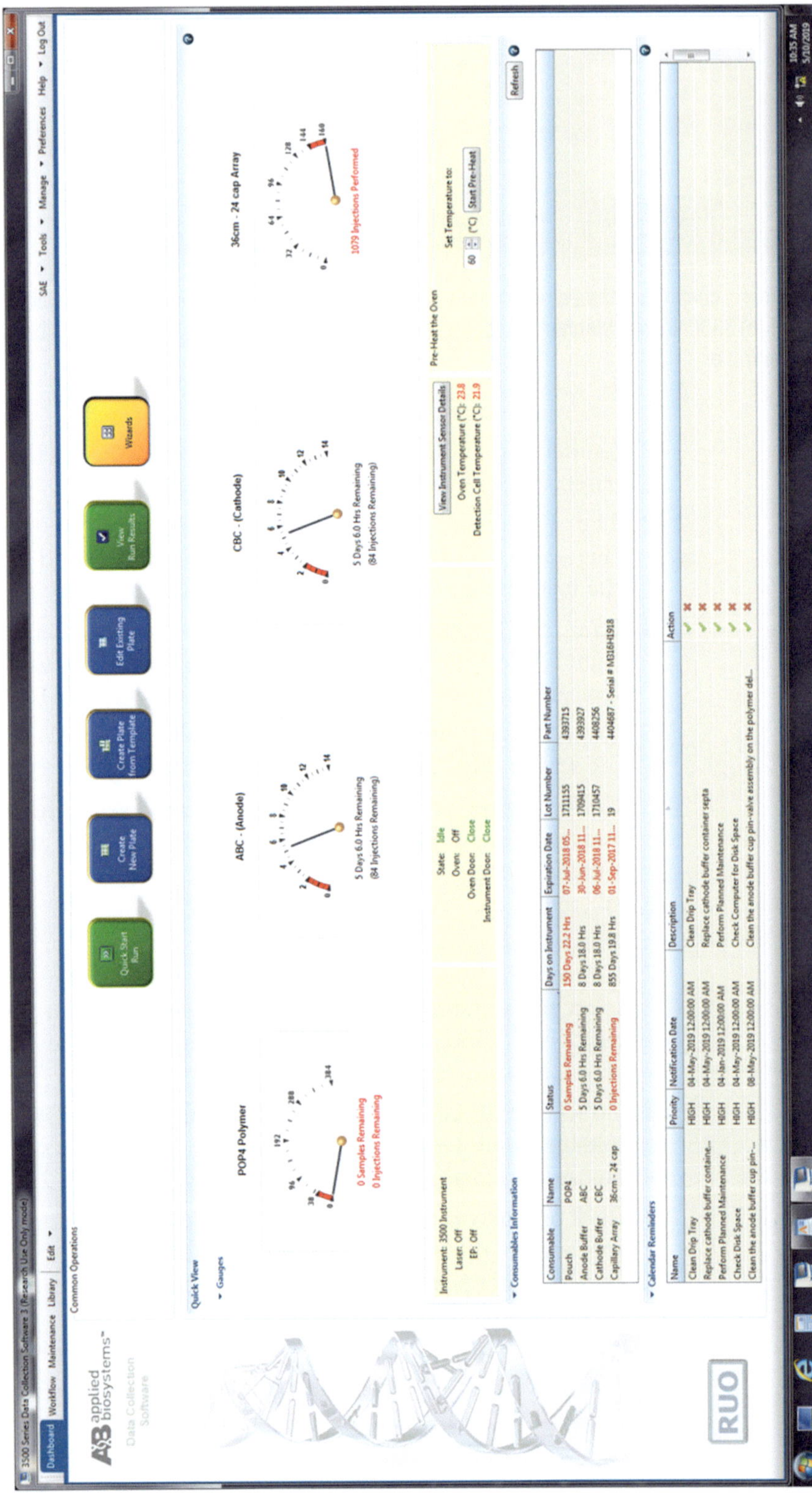

Fig. 23.6 Home screen of the data collection software

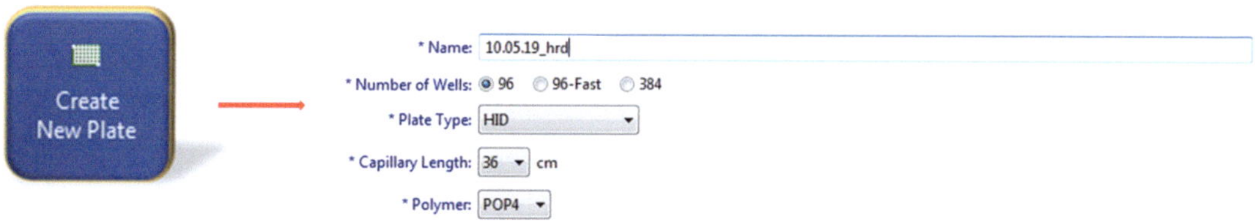

Fig. 23.7 Assignment of plate name and other attributes

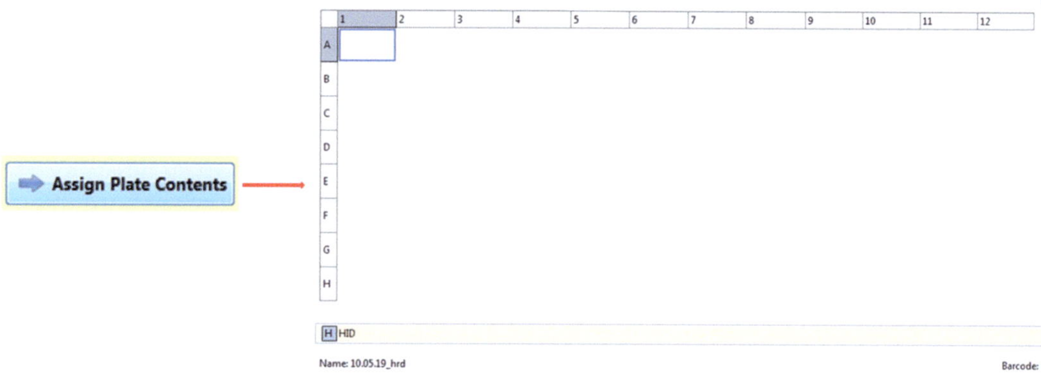

Fig. 23.8 Plate content assignment

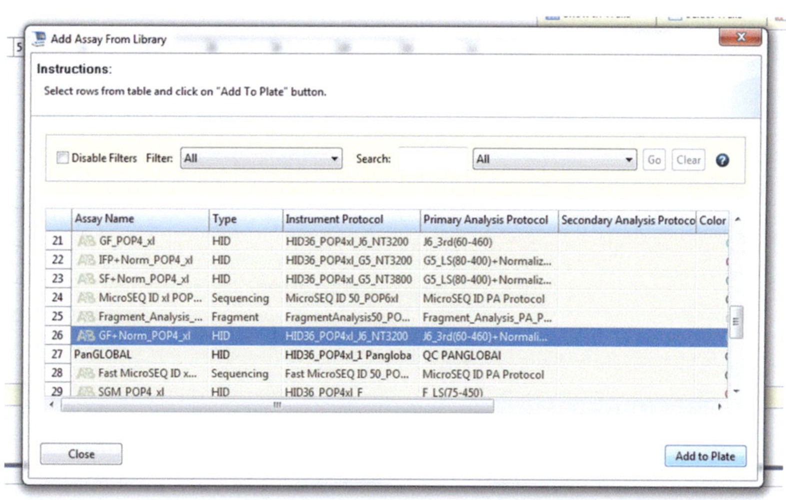

Fig. 23.9 Adding information of the assay to be performed

Similarly, select the ladder present at 3H position and assign "Allelic Ladder" from the drop-down menu (Fig. 23.14).

22. Click on "Link Plate for Run." Wait for sometime; a window will be displayed as in Fig. 23.15.

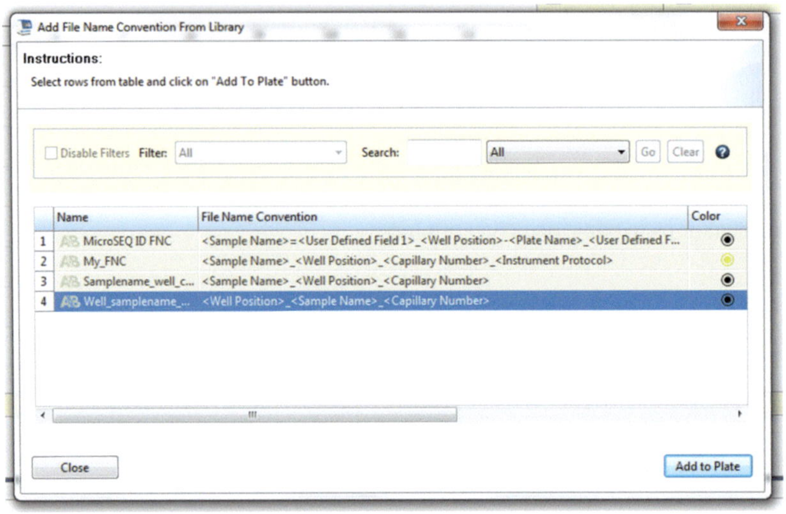

Fig. 23.10 Adding information on the File name Convention

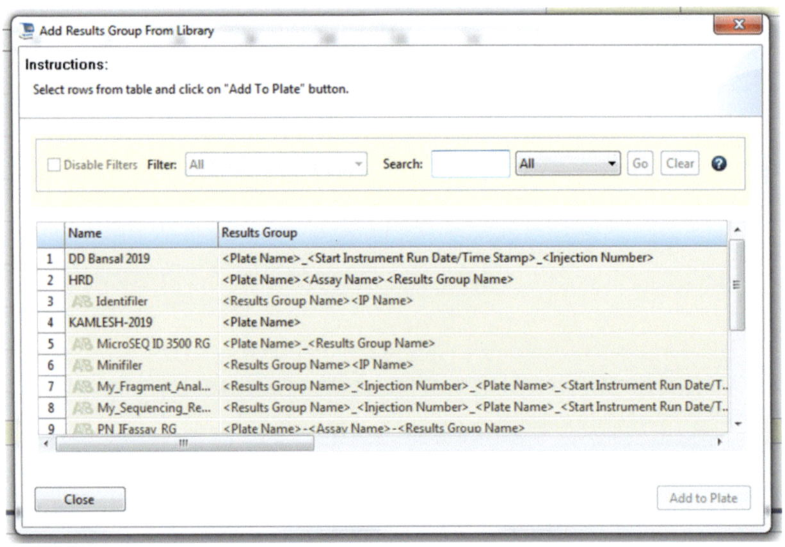

Fig. 23.11 Adding information on the "Result Group"

23. To be in safe side, click on "Preview Run." Check that all the information attributed to each well is correct (Fig. 23.16).
24. Go back to "link plate for run" and click on "Start Run" (Fig. 23.17).
25. When the run starts the green indicator of the machine blinks and the window displays as follows (Fig. 23.18).

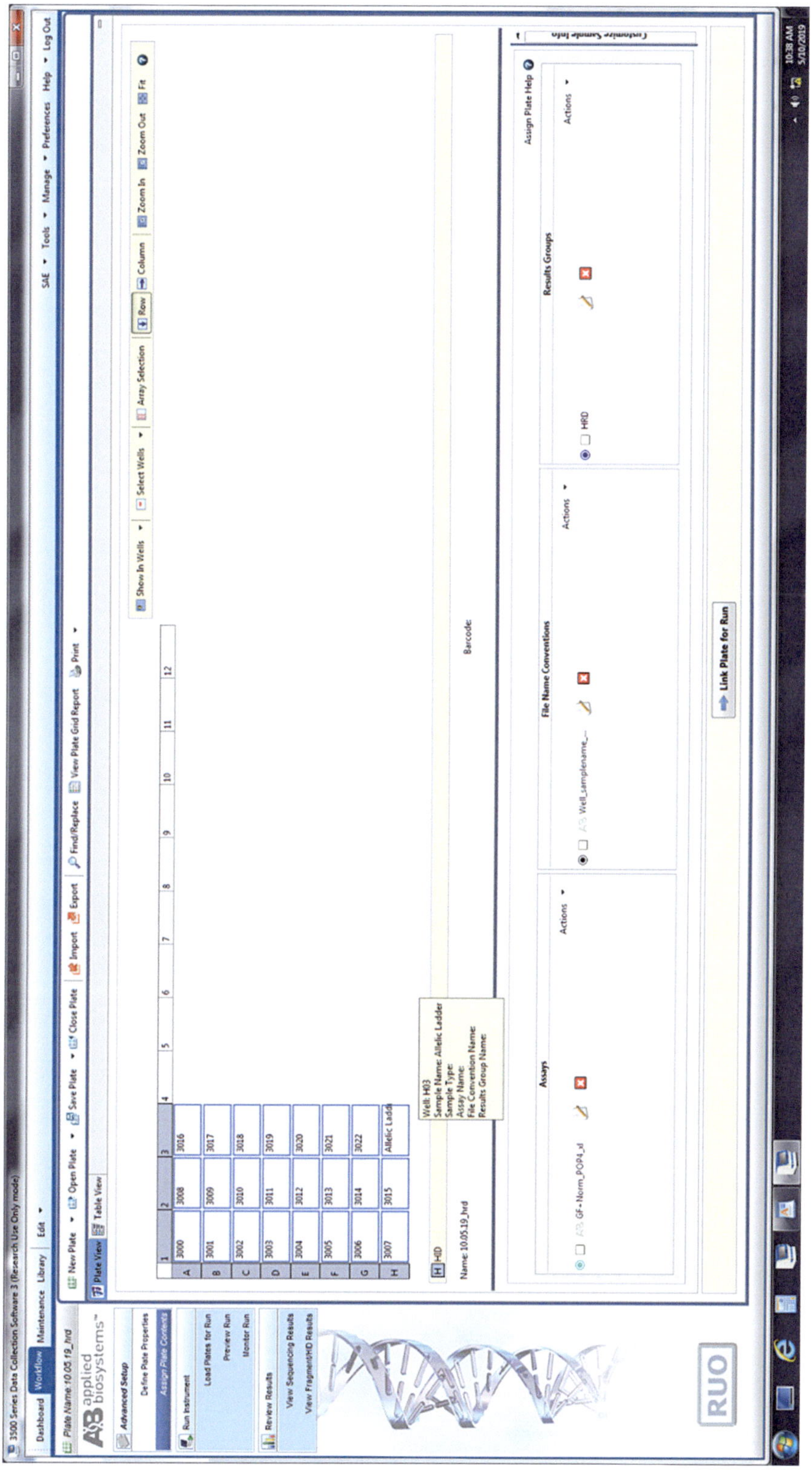

Fig. 23.12 Adding information of samples loaded at the corresponding positions of the 96-well plate

Fig. 23.13 Selection of the wells and to link with the assigned Assay, File name Convention, and Result Group

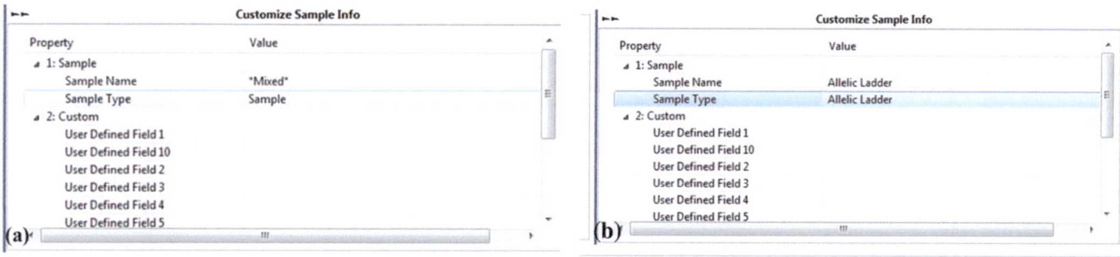

Fig. 23.14 Assigning the customized (**a**) sample and (**b**) allelic ladder position to the data collection software

26. When the run ends, blinking of the green indicator stops and the window displays as follows (Fig. 23.19).

Instrument Calibration and Maintenance

27. There are two types of calibration, i.e., spatial calibration and spectral calibration. You need to perform spatial calibration each time after you replace the capillary array, during opening of detector cell, or after moving the instrument. However, spectral calibration needs to be performed when using a new dye set, changing capillary array, and changing polymer type and during repeated observance of reduced spectral separation in raw data.
28. For spatial calibration, go to "Maintenance" and select "Spatial Calibration."
29. Select "Fill" to fill the capillary with polymer prior to calibration.
30. Select "Start Calibration." The result will be displayed as the run progresses (Fig. 23.20).
31. The spatial calibration result should be evaluated as follows: one sharp peak for each capillary, one "+" mark at the top of each peak, and all the peaks having an almost equal peak height.
32. If you are satisfied with the result, select "Accept Results" or else "Reject Results" and perform troubleshooting as described later in the chapter.
33. If accepted result is found, select "Export," enter a name of the export file, select its type as "CSV," and click on "Save."
34. Before performing spectral calibration, make sure that all the consumables are appropriate and buffer levels are up to the mark.
35. Before using Global Filer™ kit, use DS-36 Matrix Standard (dye set J6 which contains DNA fragments labeled with 6-FAM™, VIC™, NED™, SID™, TAZ™, LIZ™) for spectral calibration.

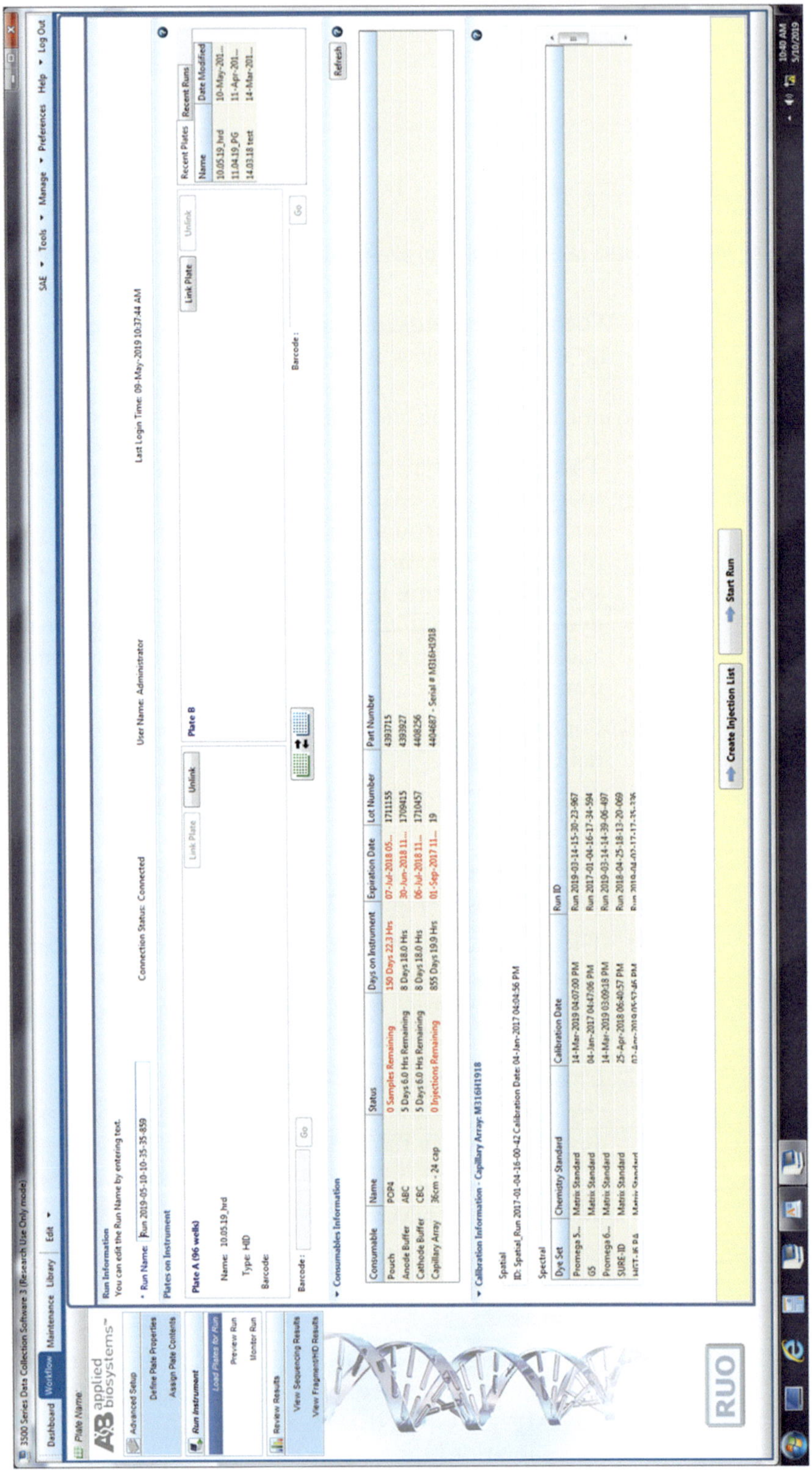

Fig. 23.15 The displayed window after linking the plate for run

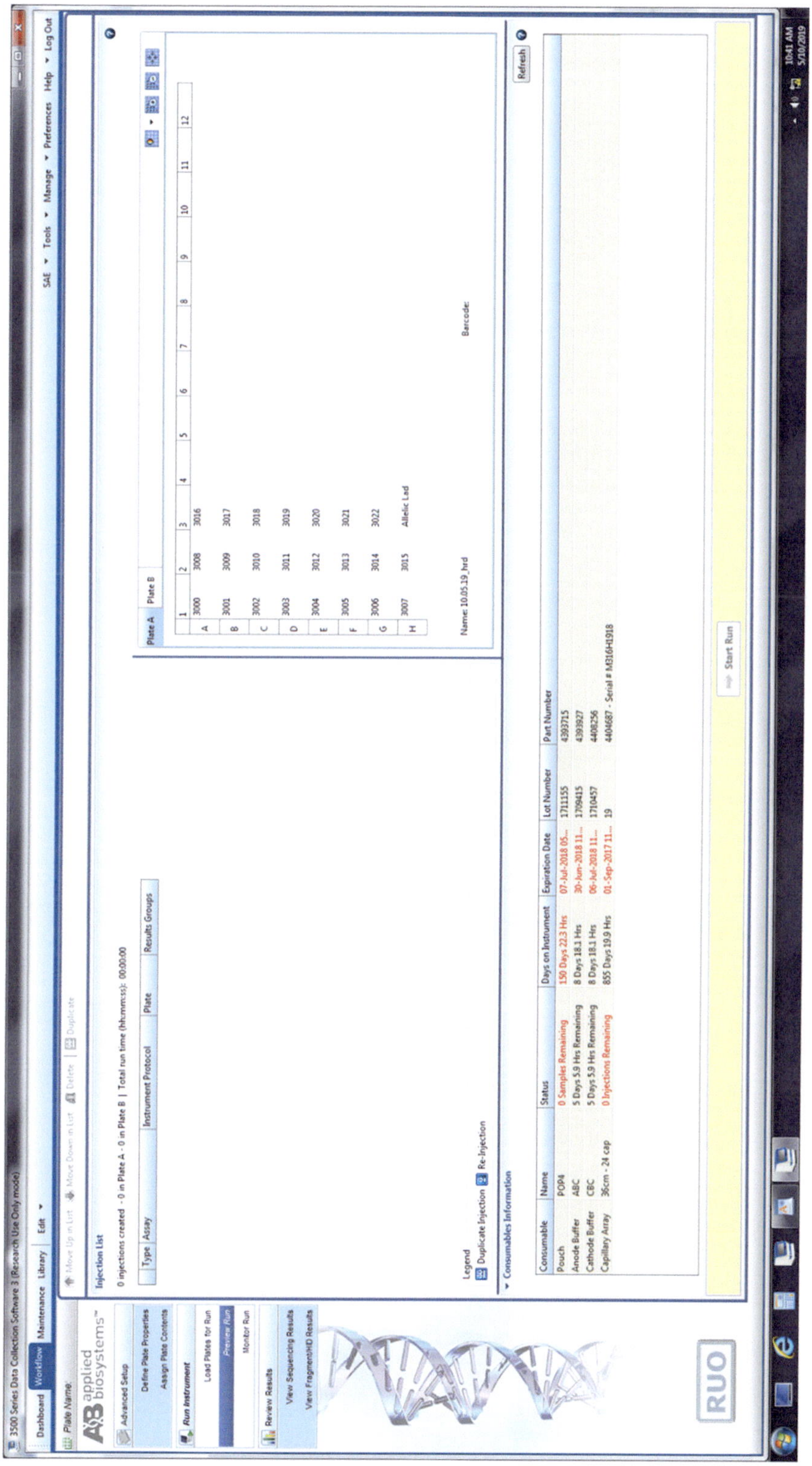

Fig. 23.16 Select "Preview Run" to assess the input of correct information to each sample

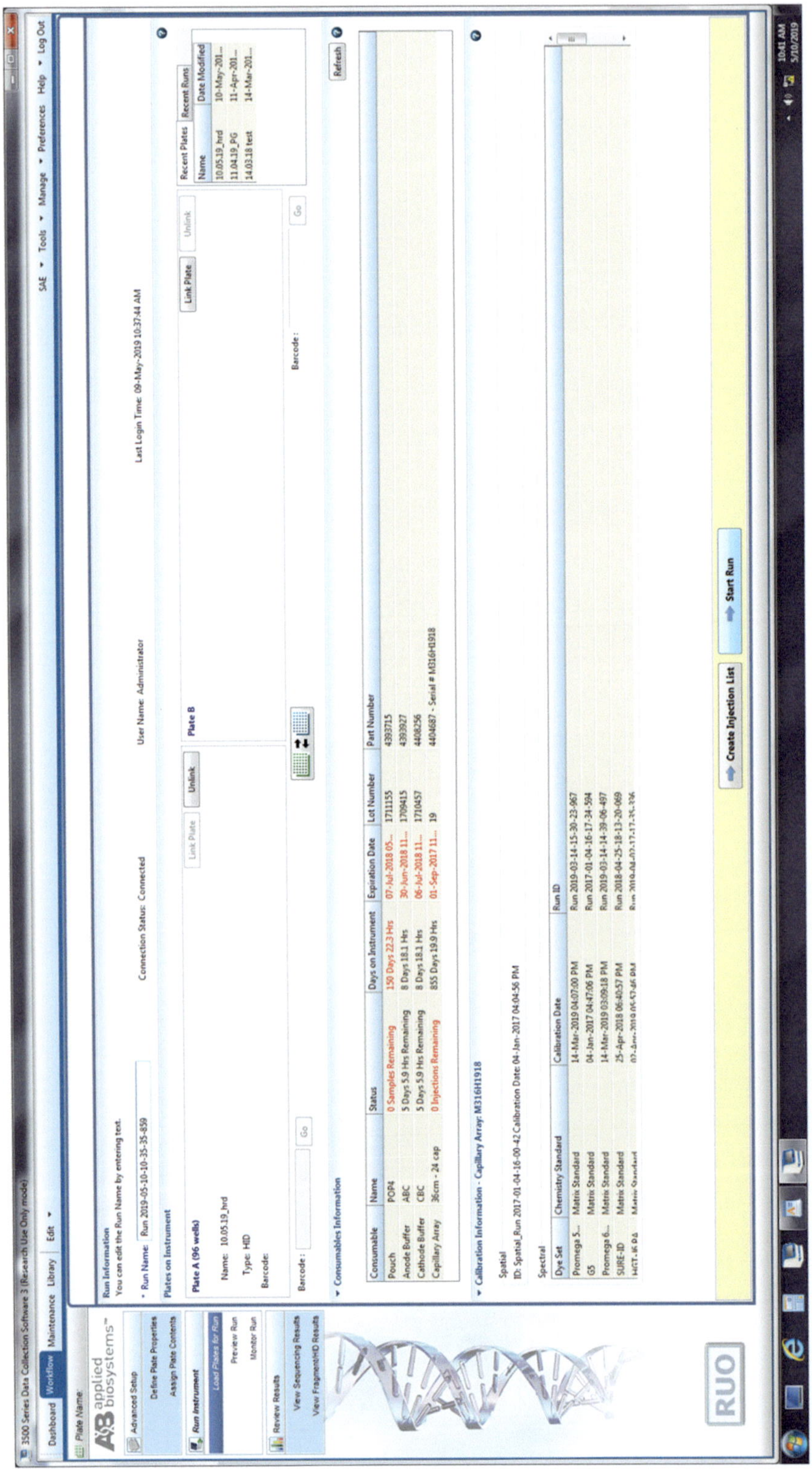

Fig. 23.17 Go back to "Link Plate for Run" and "Start Run"

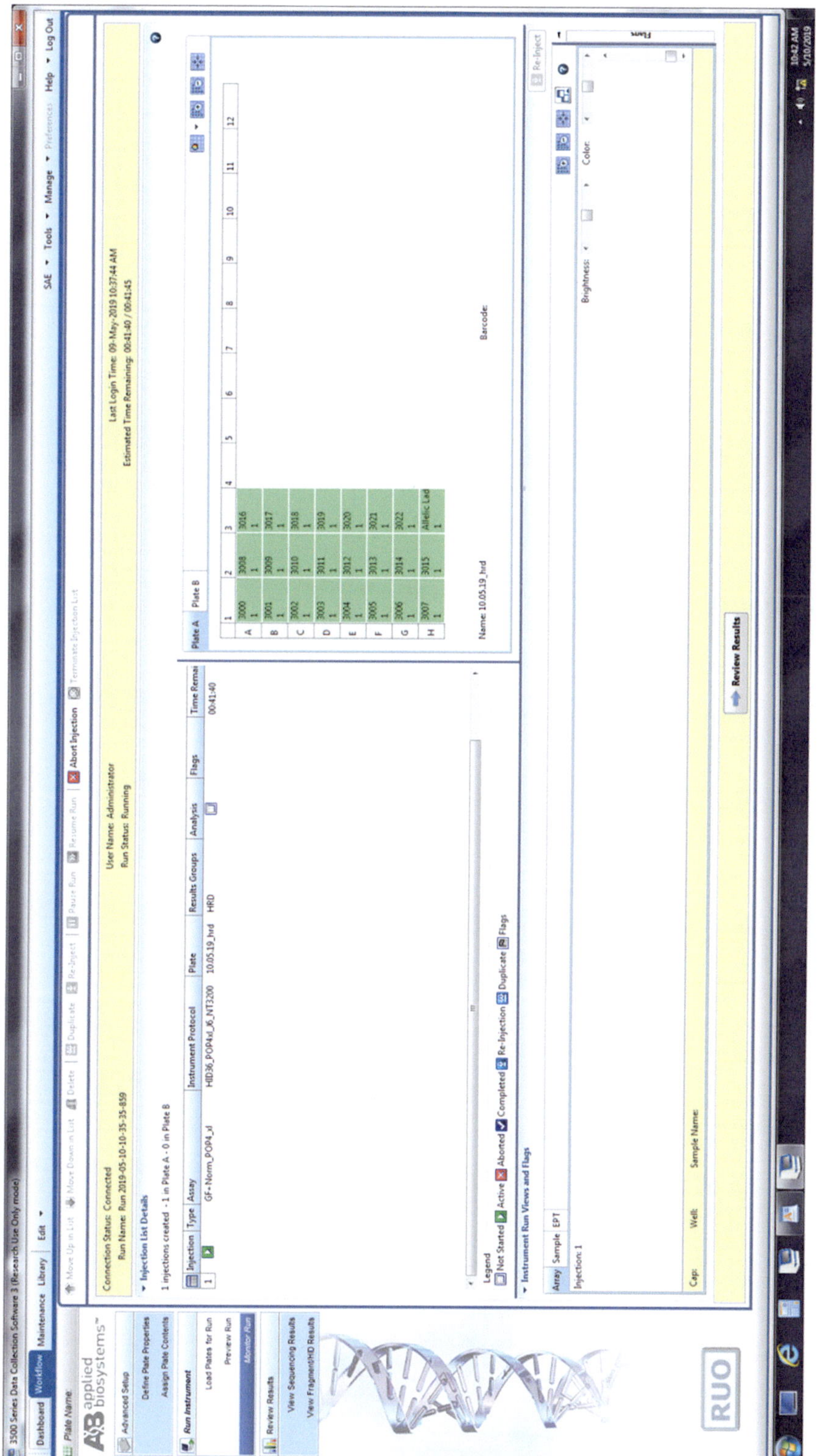

Fig. 23.18 Display of the window when the run starts

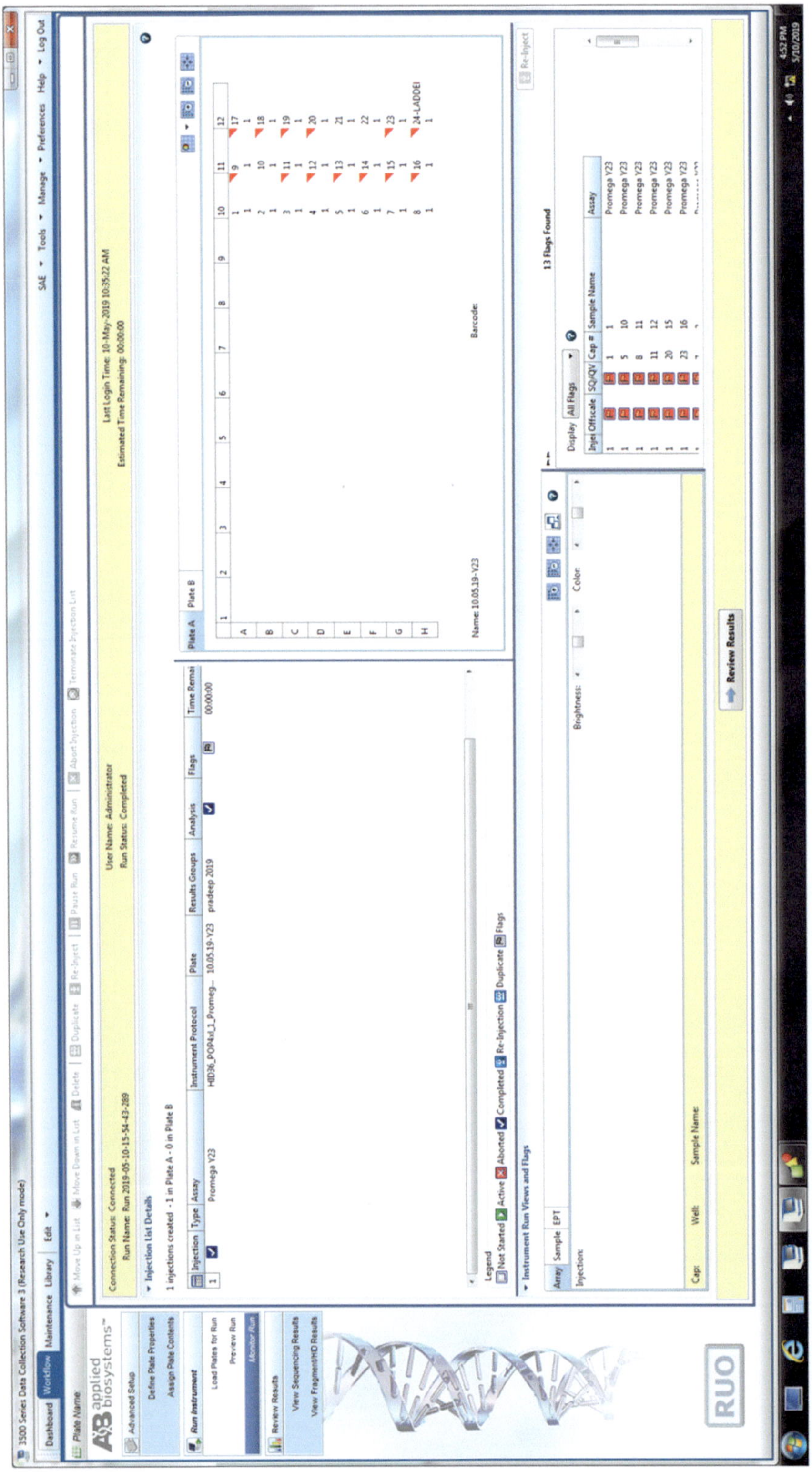

Fig. 23.19 Display of the window after completion of the run

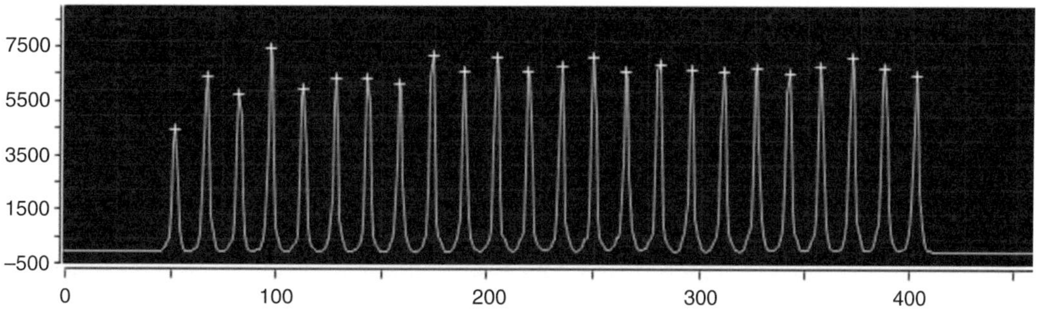

Fig. 23.20 An acceptable result of a spatial calibration in a 24-capillary system

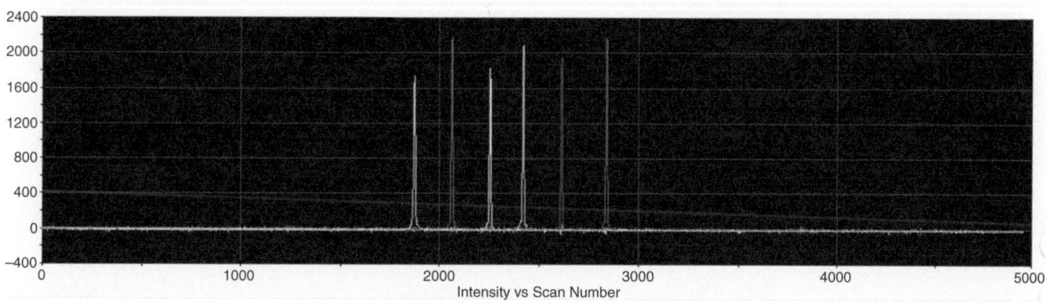

Fig. 23.21 A passing six-dye spectral calibration result

36. Vortex the matrix standard briefly and spin in a centrifuge.
37. Prepare the standard by mixing 6 μl of standard into a 294 μl of Hi-Di™ Formamide. Vortex the mixture and spin briefly in a centrifuge.
38. Dispense 10 μl of the mixture to the A1 to H1, A2 to H2, and A3 to H3 wells of a 96-well plate.
39. Cover the plate using appropriate septa.
40. Denature the plate contents at 95 °C for 5 min followed by immediate snap cool on ice.
41. Go to "Maintenance" and select "Spectral Calibration."
42. Specify the location of the samples placed in the 96-well plate in the instrument.
43. Select the standard chemistry and the dye set for which you want to perform the calibration. N.B.: Any new dye set can also be calibrated on the instrument by creating the custom dye set.
44. Select "Start Run." The system will perform three injections and the run data is displayed after the end of each injection (Fig. 23.21).
45. Examine the result for all the capillaries and the result obtained among different injections.

46. When all the capillaries yield an acceptable result, select "Accept Results" or else "Reject Results" and perform troubleshooting as described later in the chapter.
47. During HID applications, three maintenance activities need to be performed regularly, i.e., replenish polymer, change the buffers, and replace the old capillary with a new one.
48. For change in buffer, bring the anode buffer container (ABC) from the storage and equilibrate it to room temperature.
49. Ensure the presence of 1× buffer in the larger side of the container, verify the content up to the mark, and peel off the seal of the ABC.
50. Place the ABC at the designated position of the instrument below the pump. Close the instrument door.
51. Refresh the dashboard of the software to check the updated status of the ABC.
52. Similarly, bring the cathode buffer container (CBC) from the storage and equilibrate it to room temperature.
53. Verify the content of CBC up to the mark and peel off the seal of the CBC.
54. Place appropriate septa on both sides of the CBC and place the CBC on the autosampler.
55. Close the instrument door and refresh the dashboard of the software to check the updated status of the CBC.
56. When the polymer is utilized, the dashboard itself shows warning signal to replenish the polymer pouch. In such cases, the used polymer pouch needs to be replaced with a fresh pouch.
57. To replenish polymer, bring the polymer pouch from the storage and equilibrate it to room temperature.
58. Select "Replenish Polymer" from the maintenance wizards screen.
59. Open the door of the instrument and replace the old polymer pouch with a new one. Close the door and follow instructions of the maintenance wizard.
60. Refresh in the dashboard to verify the updated status of the polymer.
61. Check for the presence of any bubbles in the path. If found, follow the wizard "Remove Bubbles" which will take around 10–15 min to complete.
62. Additionally, to change the capillary array, select "Install Capillary array" from the maintenance wizard.
63. Replace the old capillaries with a set of new capillaries and close the door carefully.

64. Follow the instructions given by the maintenance wizard. Refresh the dashboard to verify the updated status of the capillary.

5 Observation

You can monitor the electrophoresis process during the run as following parameters can be visualized.

(a) In the instrument Run Views select "EPT" and for an ideal electrophoresis run the EPT graph should be found as given in Fig. 23.22.

(b) In the instrument Run Views select "Array" and the result will appear as given in Fig. 23.23. Presence of all the dye sets in each capillary suggests the separation of amplified products in that capillary.

(c) In the instrument Run Views select "Sample" followed by selecting allelic ladder or a sample; the result will appear as given in Fig. 23.24. Presence of peaks of all fluorescent colors suggests proper amplification and electrophoresis.

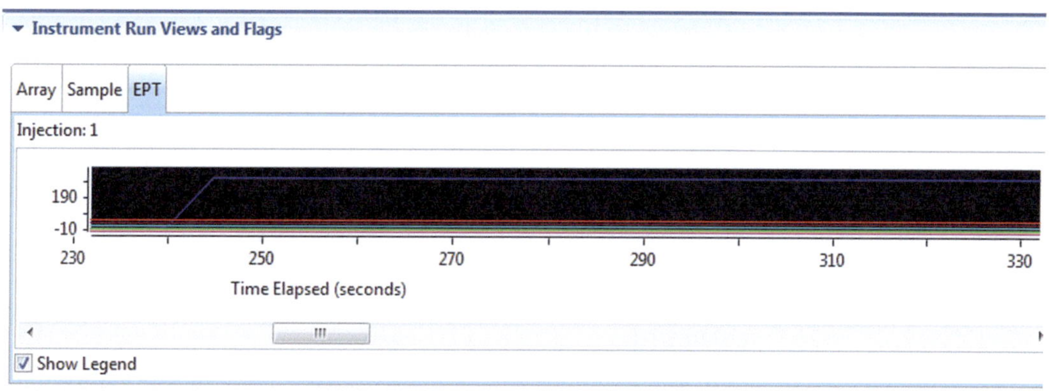

Fig. 23.22 An ideal EPT curve shown by the instrument

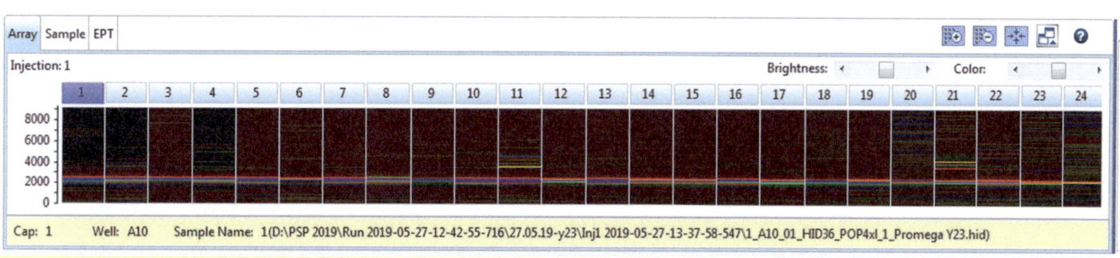

Fig. 23.23 Visualization of samples present in the corresponding capillary arrays

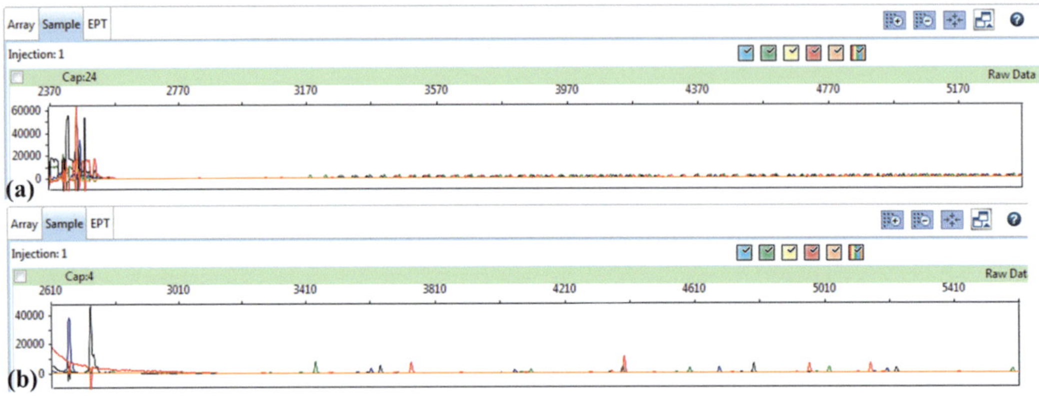

Fig. 23.24 Visualization of peaks for the (**a**) allelic ladder and (**b**) sample

6 Precautions

- Wear gloves while dealing with the reagents and plasticware.
- Always use filtered microtips or aerosol-resistant microtips to prevent any possible contaminations.
- As some of the chemicals are hazardous, they should be handled carefully.
- Always store the reagents in appropriate storage conditions.
- Store the matrix standard in dark conditions.
- Check for the presence of crystals in the polymer pouch before using it on the instrument.
- Never mix the pre-PCR and post-PCR area.
- Avoid repeated freezing and thawing of the allelic ladder and size standard.
- All the reagents, i.e., polymer, ABC, CBC, and conditioning reagents, should be brought to room temperature prior to putting them on the instrument.
- Always maintain the room temperature for optimal performance of the instrument.
- Before performing the experiment always switch on the computer first followed by the instrument and finally the data collection software.
- Before launching the data collection software always wait for sometime till the establishment of connection between the instrument and the system.
- Denature the plate just before putting it in the instrument.

- Used septa should be cleaned carefully; otherwise it may lead to contamination.
- Check the alignment of the 96-well plate before putting it in the plate base and plate retainer.
- Recheck the samples loaded in the 96-well plate and the samples assigned to the instrument for electrophoresis.
- Never use the reagents (polymer, ABC, CBC, and conditioning reagent) without proper seal.
- Never use expired reagents.
- After run is complete do not close the data collection software immediately as it may take few more minutes to complete data collection after completion of run.
- Always check the status of electrophoresis by monitoring the EPT graph prior to data analysis.
- In between runs do not close the data collection software and open it. It may hamper the life span of the laser present in the instrument.

7 Troubleshooting

Problem	Tentative cause	Possible solutions
Continuous blinking of yellow light	Pause during run	Select "Resume Run"
	Door of the instrument remained open	Carefully close the door of the instrument
	Run failure	Check for any error message and rectify it. Rerun the samples or conduct another run
"Start Run" icon is not responding	Initialization of the instrument is not complete	Wait for 10 s till the yellow light turns to green after closing the instrument door, and then proceed with "Start Run"
Run paused unexpectedly	Reading of RFID is not appropriate in the instrument	Refresh the dashboard; if the problem persists, restart the computer, instrument, and software

(continued)

Problem	Tentative cause	Possible solutions
Run paused unexpectedly with error of electric discharge	Less content in ABC	Replace with a fresh ABC
	Leak in the system	Run the Bubble Removal wizard and rerun
	Polymer crystals present in the buffer pin valve	Ensure cleaning of the buffer pin valve at regular interval
Run stopped after one or more than one injections	RFID of the consumables could not be read by the instrument	Check the connection between the instrument and software and proceed with a new injection
The dashboard does not update automatically after replenishing the consumables	The wizards have not been refreshed	Refresh the dashboard or start a new run
Pre-run check validation failed	Check the dashboard for the status of the consumables	Replace the ABC/CBC/polymer/capillary for which the pre-run check validation failed
Error during linking the plate	No plate in position A	If you have loaded the plate at position B, link the plate with position B
	No plate detected	Place the plate at position A and follow the instructions
During spatial calibration "start" button is disabled	No communication between instrument and data collection software	Restart the system, instrument, and software; then perform the spatial calibration again
During spatial calibration unusual peaks or no peaks appear	Detection cell is not installed properly	Reinstall the capillary array, followed by fill array, and perform the experiment again
	Stability of the instrument not achieved	Repeat the experiment after sometime
	Capillary is broken	Check the capillary; if found broken replace it

(continued)

Problem	Tentative cause	Possible solutions
Bad spatial calibration results	Faulty capillary array	Replace the capillary array with a new one and perform the experiment
Message pops out as "spatial calibration error"	Conditioning reagent is present in place of polymer	Replace the conditioning reagent with polymer pouch
No signal during spectral calibration	Sample is not prepared properly	Prepare the samples again with fresh Hi-Di™ Formamide and perform the experiment again
	Bubbles present in the wells containing samples	Remove the bubbles by centrifugation and perform the experiment again
	Tip of the capillary is not touching the samples properly/capillary array is not properly installed	Check the volume of the sample loaded; if recommended volume is applied, call the service personnel of the instrument
"No spectral files found" error is shown	Capillary is blocked	First try with refilling the capillary array; if the problem persists install a fresh capillary array and perform the experiment again
	Incorrect run module/ chemistry file/dye set selected	Correct the inappropriate files and perform the experiment again
	Expired matrix standard used	Check the matrix standard and if necessary use a fresh set of reagents
Peaks found below the required amplitude	$\geq$1 peaks found below the recommended amplitude, i.e., 750	Rerun the spectral standards with a higher amount of standard

(continued)

Problem	Tentative cause	Possible solutions
Spikes appeared in the result	Faulty polymer	Replace the faulty polymer with a new fresh polymer pouch and perform the experiment again
	Contaminants/crystals found in the polymer	
	Bubbles present in the polymer path	Perform the bubble removal option using the wizard and perform the experiment again
Baseline is elevated	Spectral calibration is poor	Perform the experiment again

Chapter 24

Analysis of Capillary Electrophoresis Results by GeneMapper® ID-X v 1.5 Software

Objective: To analyze the capillary electrophoresis results by using GeneMapper® ID-X v 1.5 software.

1 Introduction

Data interpretation is the most crucial step of forensic DNA analysis and needs enormous expertise to deal with it. As forensic DNA typing is a comparative assessment of results between the control/reference sample and the test/questioned sample, three possible outcomes may appear for an experiment, i.e., a match, a mismatch (maybe mutational), or an inconclusive result. All these possible results should be evaluated carefully by an examiner. With the advancement of technology, many software have eased the work of the forensic scientists; however, thorough examination of a true allele designation can reverse the opinion of a case examined through the technique.

Multiplex PCR followed by capillary electrophoresis generates a huge data that cannot be analyzed manually. Data interpretation is the most laborious step of STR typing. Due to complex nature of the data generated during capillary electrophoresis, data interpretation requires 50% or more of the resources. Thus, to minimize human intervention during data analysis, many software are in practice nowadays (for comprehensive list see Appendix B); the GeneMapper® ID-X is widely used across the globe in forensic DNA analysis. Besides designating alleles to the amplicons generated in the form of peaks, the software is also capable of distinguishing between a true peak and the possible artifacts, for example dye blobs, stutters, and spikes. Though GeneMapper® is capable of generating a DNA profile by analyzing the capillary electrophoresis data, it still requires manual review. The process quality values

H. R. Dash et al., *Principles and Practices of DNA Analysis: A Laboratory Manual for Forensic DNA Typing*, Springer Protocols Handbooks, https://doi.org/10.1007/978-1-0716-0274-4_24, © Springer Science+Business Media, LLC, part of Springer Nature 2020

(PQVs) increase the confidence of the allele designation and troubleshooting efforts. The advanced version of GeneMapper® provides solution to analyze data from single-source samples as well as for a mixture interpretation.

2 Principle

After completion of electrophoresis, the Genetic Analyzer 3500 system generates .hid (human identity) files that contains electronic information on the laser power and run current besides information of the amplified DNA fragments. The overall role of data analysis software is to convert the time/scan points to the DNA sizes with reference to the internal size standard encrypted in the electronic file (.hid file). Further by the utilization of kit-specific bins, panels, and allelic ladders the amplified fragment is designated as an allele. When the software generates a DNA profile, analysts' role becomes imperative to assess the electrophoretic data as a noise or a proper peak, distinguishing a proper peak from common artifacts, pairing alleles to form a genotype, and combining the genotype of all the loci tested to generate a complete DNA profile. Prior to analysis of a DNA profile the following terms associated with result interpretation should be understood carefully to establish sensitivity, reproducibility, and precision of the obtained data as well as correct assignment of heterozygosity and assessment of mixture samples.

Thresholds: During low-copy-number DNA analysis, it becomes highly difficult to distinguish between a true peak and the technical artifacts. Thus, peak height thresholds have been established to designate a true allele. In such cases, when the relative fluorescent unit (RFU) value of peak height exceeds the designated threshold value the peak will be called as an allele (Fig. 24.1). Though there is no standard rule for establishing a threshold value, it should be a part of any laboratory's internal validation procedure. It can be maintained by observing the signal-noise ratio or the software sets it arbitrarily. Though the software manufacturer (Thermo Scientific) recommends the peak height threshold to be 150 RFU, forensic samples as low as 50 RFU can be used for allele calling.

An ideal data: Unlike "bands" of slab gel electrophoresis, the results of capillary electrophoresis are generated in the form of "peaks." A proper peak should start from the baseline and finish at baseline. At the signal level, the top of the peaks should be well above the analytical threshold and well differentiated from the noise. The loci should have well-defined peaks within the pre-designated shaded beans defined by the allelic ladder. In an ideal profile, all the peaks for homozygous alleles and all the peaks for heterozygous alleles should have the same peak height ratio. The peak height of the homozygous allele should be twice that of the peak height of a heterozygous allele (Fig. 24.2). An ideal profile

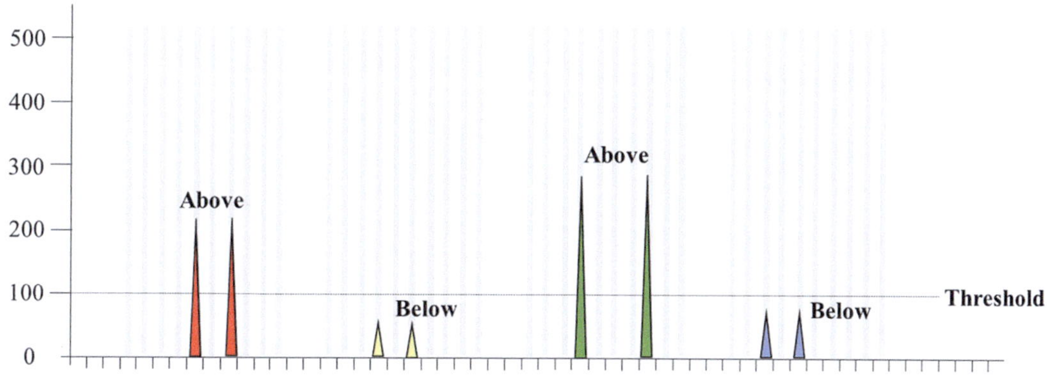

Fig. 24.1 Role of threshold in allele calling for a sample

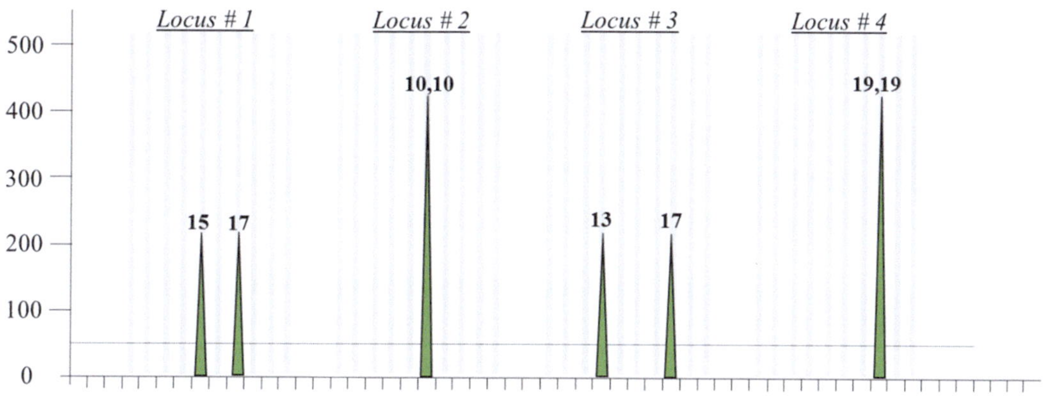

Fig. 24.2 An ideal data of alleles from four genetic loci

without any artifact is capable of differentiation of contributor's DNA profile from a mixed sample and low DNA sample analysis.

Spikes: Spikes are the known artifacts that appear due to voltage fluctuation during electrophoresis. The fluctuation in voltage generally arises due to the presence of minute air bubbles in the capillary or crystal formation in the polymer or contamination of polymer by the fluorescent materials. More amounts of spikes are generally visible in the raw data than the analyzed data. Spikes when not appearing along the true peak generally appear at the same data point for all the dye sets. They also do not fall as per the true peaks as they do not have a broader base (Fig. 24.3). The appearance of spikes in the result can be minimized by using a fresh set of reagents and samples. However, the frequent occurrence of spikes is an indication of system malfunctioning.

Dye blobs: Dye blobs can be clearly distinguished from the true peaks as they have a broader base and are mostly seen in a specific dye set for one analysis (Fig. 24.4). Dissociation of fluorescent dyes from the labeled primers due to improper storage conditions

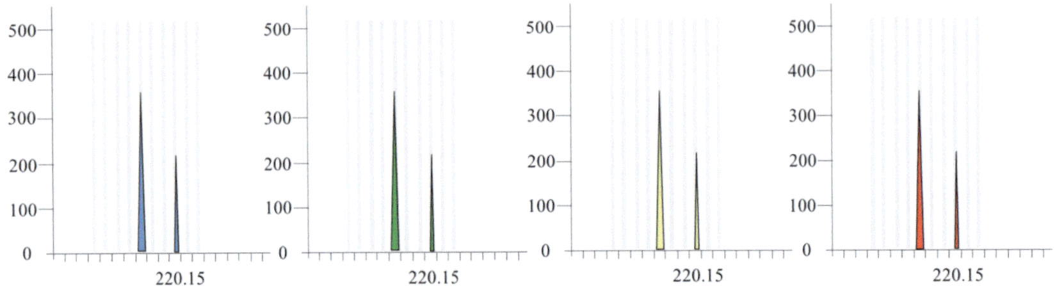

Fig. 24.3 Occurrence of spikes at the same data point in all the dye sets

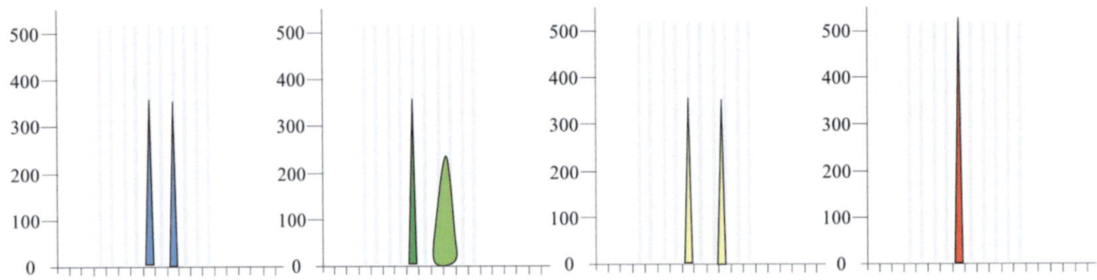

Fig. 24.4 Occurrence of dye blob with wider base in the green dye set only

generates dye blobs in an electrophoresis data. To minimize the dye blobs in result, the primers should be stored in the recommended storage conditions. Further, purification of the amplified product using a filtration unit will improve the result by passing through the dissociated primers and retaining the amplified DNA fragments. However, in a set forensic laboratory system and use of standardized protocols the filtration step may not be required.

Noise: It is the series of baseline peaks which are non-reproducible and present as backgrounds to the sample results. Noise appears due to electronic fluctuations, presence of air bubbles and crystals in the polymer, and low-level sample contamination. Noise does not create problem in the analysis as they are present in the baseline. However, if they appear close to the threshold value they can be misinterpreted as a true peak. The most useful way to distinguish between noise and a true allele is that noise is not reproducible whereas peaks are always reproducible. When the result is confusing between the allele and noise, the sample should be subjected to electrophoresis again and the result can be interpreted.

Pull-up: Pull-up is a major problem when high amount of template DNA is added to a PCR setup followed by electrophoresis. The oversaturated data pulls up one color dye to another dye set and in such cases the software is unable to distinguish between different colors. Re-amplification with less amount of input DNA can minimize the pull-ups. Careful observation of the pull-ups can

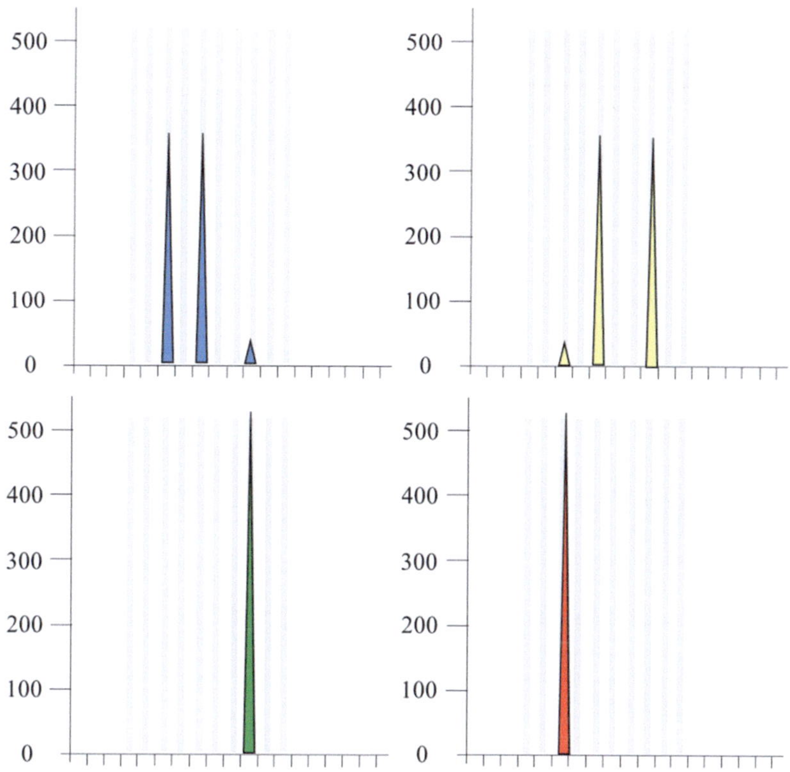

Fig. 24.5 Occurrence of pull-ups in the blue and yellow dye sets due to oversaturated peaks present in the green and red dye sets

distinguish it from the true peaks as they appear at the above or below dye set at the same position of a saturated data (Fig. 24.5). Proper quantification and addition of optimum quantity of DNA in a PCR process will minimize these artifacts; however if the problem persists even after adding optimal DNA quantity, a new matrix should be run and applied to the set of results.

Stutter: Stutter is a PCR artifact which is formed as one repeat shorter or larger than the true PCR product and accumulates as a minor product. Though the mechanism of stutter formation is not completely understood, the bulging of the template strand during strand breathing results in the formation of a reverse stutter whereas bulging of the newly synthesized strand results in the formation of a forward stutter (Fig. 24.6). Due to its specific appearance and size, stutters can be easily identified in a single-source sample. However, it is difficult to distinguish between a true allele and a stutter in a mixed-source sample. Studies have shown the occurrence of a higher percentage of stutters in low unit repeats, homogeneity of repeat units, and large alleles. Sometimes stutter products are also helpful to distinguish between the true allele from other artifacts such as pull-ups and spikes as only PCR products contain the stutter peaks.

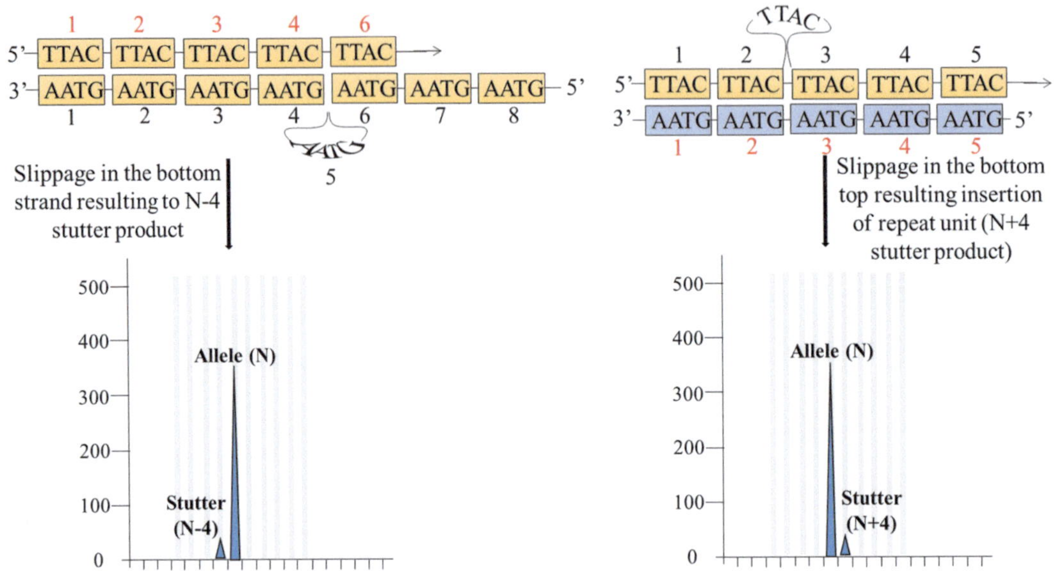

Fig. 24.6 Formation of N−4 and N+4 stutter products during in vitro DNA replication

3 Non-template Addition

It is a known PCR artifact which is generated by the *de novo* action of Taq polymerase. Taq polymerase adds nucleotides, mostly adenine, to the 3′ end of the PCR product resulting the PCR product to be one base pair larger than the actual product. The final incubation step at 60 or 72 °C favors the non-template addition by PCR process. During the final incubation step, the degree of adenylation is dependent on the sequence of the template strand. Amplification process is always carried out from the 5′ end. When the forward primer, labeled with a fluorescent dye, is used for amplification, then only the top strand/amplified fragment is detected by its fluorescence in the instrument (5′–3′). Additionally, when this labeled amplified fragment is used as template, primer binds to the 3′ end of the fragment and the terminal nucleotide is dependent on the 5′ end of the reverse primer. Studies have also revealed that complete addition of adenine by polymerase is favored by the presence of guanosine at the 5′ end of the primer. Thus, different loci possess different adenylation properties due to differences in their primer sequences. From the analysis point of view, adenylation leads to –A peaks or +A peaks, where the peaks become broader and a split peak appears when the detection system has a poor resolution (Fig. 24.7). To minimize split peaks arising due to non-template addition, the final extension step of PCR process should be optimized so that all amplicons result in the same length.

Off-ladder (OL) alleles: Human population may contain some rare alleles that are different from the common alleles of the genetic

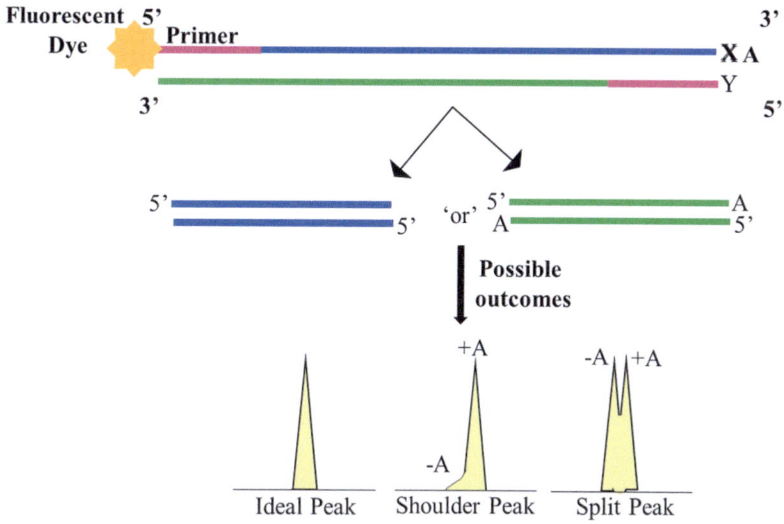

Fig. 24.7 Addition of non-template A by Taq polymerase at the end of the terminal nucleotide (X) at the 5′ end which is detected by the terminal nucleotide (Y) of reverse primer; this difference in adenylation pattern leads to the formation of –A or +A form of the amplified products

markers tested. While using any commercially available kit, the corresponding allelic ladder contains a mixture of commonly occurred alleles. When amplification of a rare allele occurs the peaks fall in between the alleles of the ladder and this is designated as off ladder (OL). Allele designation of an OL should be carefully assigned a unique allele or position. In all the instances, the off-ladder peak should be confirmed by re-PCR.

Microvariant alleles: It is an unusual finding where a microvariant represents an incomplete repeat of a STR marker allele. When an allele is designated as 15 for D18S51, it represents 15 repeats of AGAA. However, when a microvariant allele appears as 14.2, it denotes 14 complete repeat units of AGAA followed by a partial repeat of two nucleotides. Most of the commonly occurring micro-alleles are included in the allelic ladder; hence a microvariant allele is called during analysis. In general, a more polymorphic marker possesses a higher number of microvariant alleles such as FGA, D21S11, and D18S51.

Heterozygous imbalance: For stochastic effects such as low DNA samples, degraded DNA, and mixed samples, heterozygous imbalance is a major concern. If less amount of DNA is introduced to a PCR process, the heterozygous samples are amplified differentially. Similarly, for mixed samples, the major contributor generates a higher peak in comparison to the minor contributor. In such cases, it becomes highly important to distinguish between the heterozygote alleles and mixed alleles. It becomes imperative to calculate the heterozygosity of alleles and when the heterozygosity is found to be less than 70% it is an indicative of mixture or stochastic effect.

Dropout: It may be a single-allele dropout for a given locus or complete locus dropout when both the alleles of a locus are missing. Dropout occurs due to the stochastic effects when low DNA template or potential PCR inhibitors are present in the sample. In some instances, mutation in the primer-binding site may result in allele dropout by non-amplification. Sometimes allele sizes are outside the normal calling range of the locus and remain undetected resulting in the dropout.

Tri-allelic pattern: In some instances, three alleles can be observed at a locus of autosomal STR DNA profile in a single-source sample, which are generally the copy number variants. Due to the presence of an extra chromosome fragment or whole chromosome as in case of Down's syndrome three alleles appear in this particular locus. Two types of tri-allelic patterns have been established, i.e., type 1 where the sum of peak heights of two alleles is equal to the peak height of the third allele. Type 1 tri-allelic pattern is highly common in occurrence in comparison to type 2 where peak heights of the three alleles are balanced in nature (Fig. 24.8). In some cases, false tri-allelic pattern may be observed when an extreme off-ladder peak from an adjacent marker is misunderstood. Additionally, sometimes type 1 tri-allelic pattern may be masked by high stutters which should be taken care of during data interpretation.

Mutations: Mutation refers to the alteration in genetic complexion of an individual than that of its parents. Mutational events during meiosis can complicate the paternity testing of a disputed child. Thus, mutational events should not be confused as exclusion and need to be ascertained prior to reporting of any forensic case.

4 GeneMapper® IDX Software

GeneMapper® IDX software (Thermo Scientific) is capable of processing *.fsa* data generated from Genetic Analyzers of 31xx series, 310 series, and 37xx series instruments as well as *.hid* data generated from 3500 instruments series. It possesses an upper hand over the other data analysis software available in the market by providing improved data interpretation. Other advantages of this software include minimized edits of pull-ups, first-pass success rate of analyzed data, and efficient transfer of data for probabilistic genotyping. Flexibility in plate loading, streamlined workflow, support for languages other than English, and capability of analyzing six-dye results are added advantages of using this software.

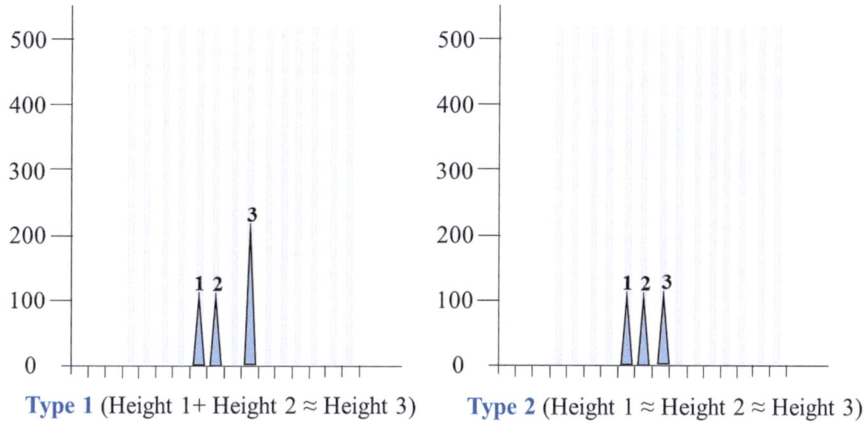

Fig. 24.8 Two types of tri-allelic pattern commonly occurred in population

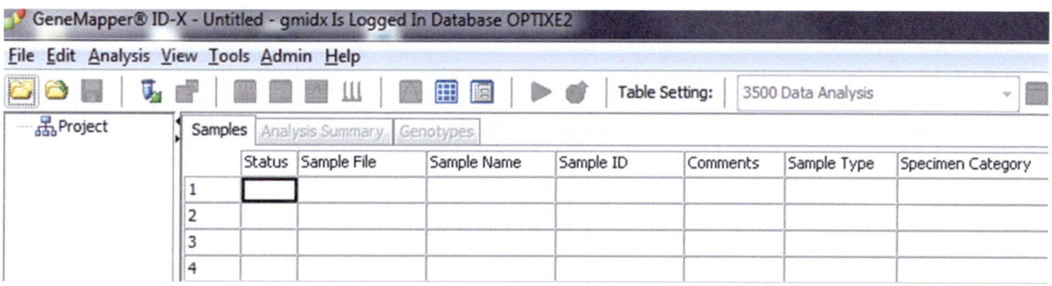

Fig. 24.9 Home screen of GeneMapper® ID-X software

5 Procedure

1. Open previously installed GeneMapper® ID-X v 1.5 in the computer with your customized user name and password. After opening the home screen will appear as in Fig. 24.9.
2. Go to "Add Samples to Project"; select the location of the results (in the previously assigned result group during plate setup), select the run, and select "Add to Project" followed by "Add" (Fig. 24.10).
3. After adding samples to a new project, the project will appear as in Fig. 24.11.
4. Select the suitable "Analysis Method," "Panel," and "Size Standard" from the drop-down menu (Fig. 24.12).

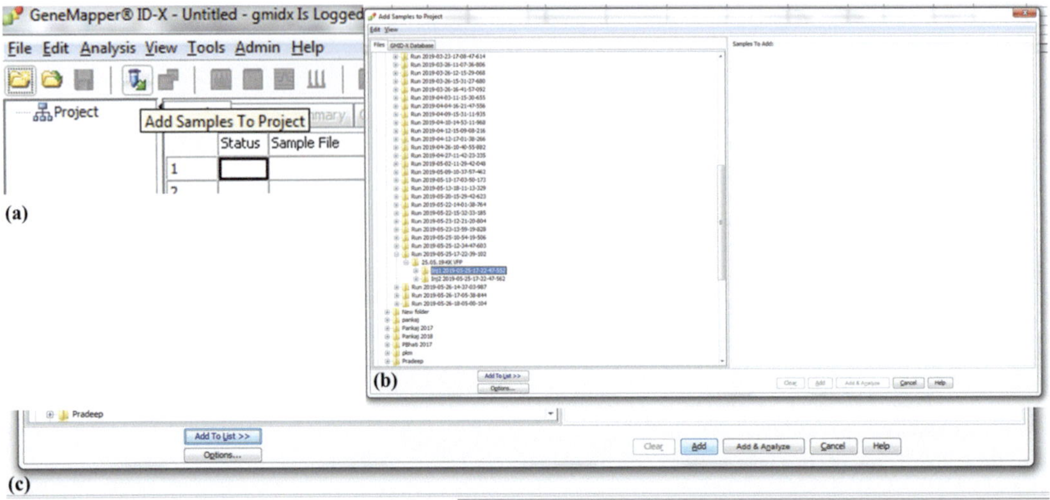

Fig. 24.10 (**a**) Adding of samples to the project from (**b**) the location of the result group for analysis and (**c**) selecting 'Add to project'

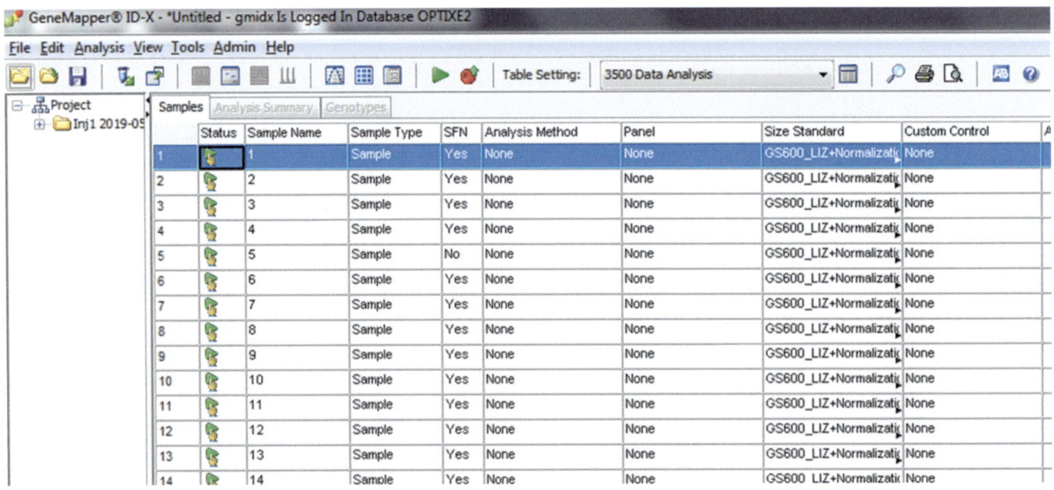

Fig. 24.11 Samples added to the project

5. Select the suitable "Analysis Method," "Panel," and "Size Standard" for all the samples, controls, and allelic ladder. You can select column for all the parameters and press "Ctrl+D" to fill all the columns with the selected "Analysis Method," "Panel," and "Size Standard" (Fig. 24.13).
6. Select "Analyze," assign project name, and select "OK" (Figs. 24.14 and 24.15).

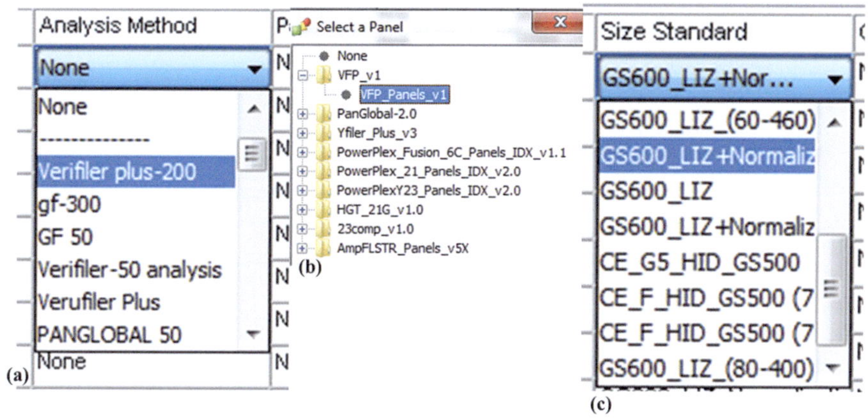

Fig. 24.12 Selection of suitable (**a**) analysis method, (**b**) panel, and (**c**) size standard for data analysis

Fig. 24.13 Selection of "Analysis Method," "Panel," and "Size Standard" for all the samples

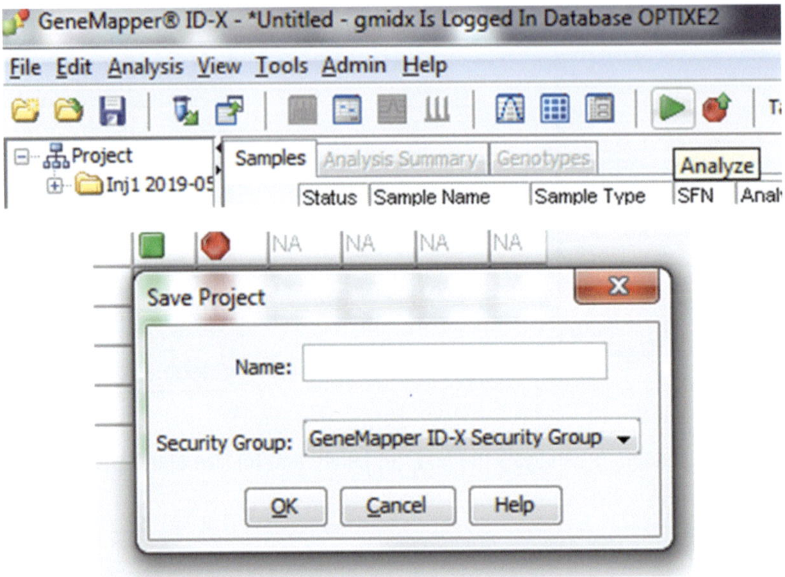

Fig. 24.14 Steps showing "Analyze" and assignment of a project name

7. After completion of analysis, the analysis summary appears (Fig. 24.16). Check for the status of quality control for allelic ladder and all the samples. When one or more threshold is not met for the allelic ladder it fails (shows the red signal besides it) and further analysis of the samples cannot be performed. Similarly, the samples and controls should be monitored for quality control before analysis.
8. Select "View" and choose "Raw Data" from the drop-down menu (Fig. 24.17). The raw data should be checked for the allelic ladder as well as the samples. Peaks from all the dye set should be present in the raw data if amplification and electrophoresis are proper.
9. Select "View" and choose "EPT data" from the drop-down menu (Fig. 24.18). The EPT graph should always appear as given in Fig. 24.19. Any deviation of EPT graph from the given figure is suggestive of an improper electrophoresis.
10. Select "Tools," choose "Size Match Editor" from the drop-down menu, and select "Size Calling Curve" (Fig. 24.19). The size calling curve should always be a straight line with the R^2 value close to 1.
11. After analyzing all quality parameters, select "View" and "Samples." Select Allelic Ladder and go to "Display Plots." The plots of allelic ladder will appear as in Fig. 24.20.
12. Similarly select any sample and go to "Display Plots" to visualize the sample plots (Fig. 24.21). Select Panes to "6" so that all the plots of a six-dye chemistry kit appear in a single window.

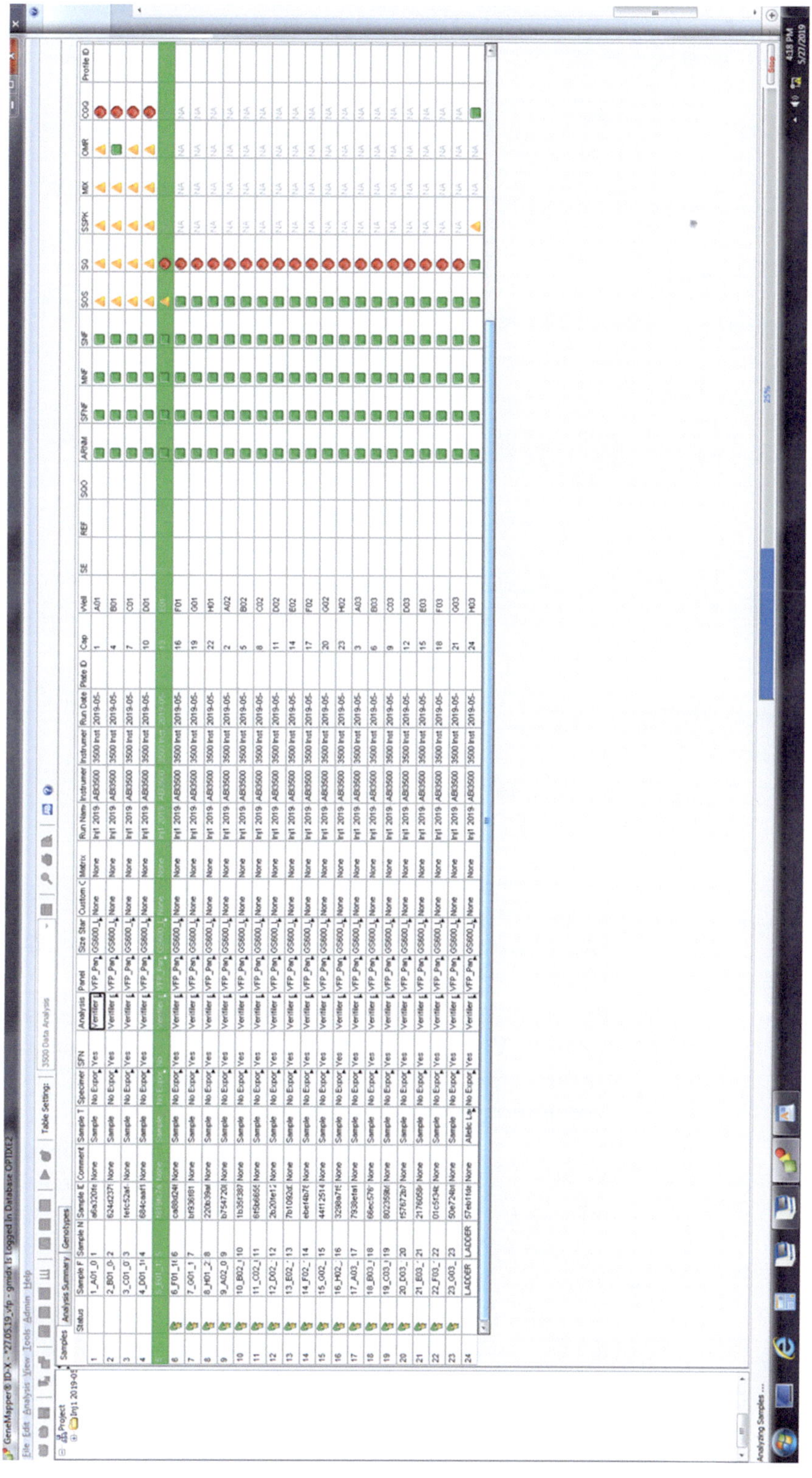

Fig. 24.15 Display of software during analysis of project

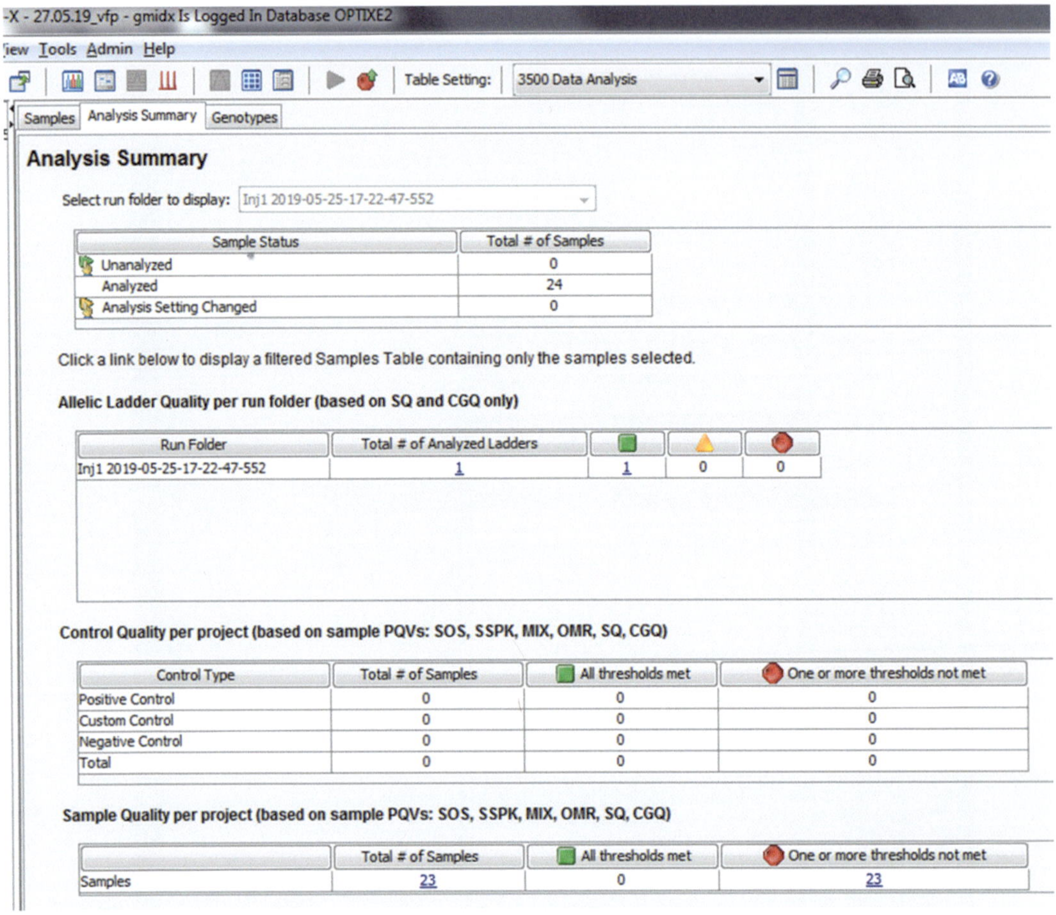

Fig. 24.16 Window displaying the analysis summary for the allelic ladder as well as the samples

Here the peaks in the orange dye denote the fragments of size standard.

13. For autosomal STR analysis the amelogenin marker shows either "XY" for female or "X" for females. Other autosomal markers either show two peaks (for heterozygosity) or one peak (for homozygosity) for a single-source sample.
14. Look for the presence of artifacts if any (as described in Sect. 2), in any of the markers. Two common artifacts, i.e., off-ladder (OL) and outside marker range (OML), are shown in Fig. 24.22.
15. After identification of an artifact, correct the profile by deleting the artifact through selecting "delete label" and giving appropriate reason for deleting the same. Figure 24.23 shows correction of a common artifact off ladder in the profile while Fig. 24.24 shows the correction of another common artifact outside marker range in the profile.

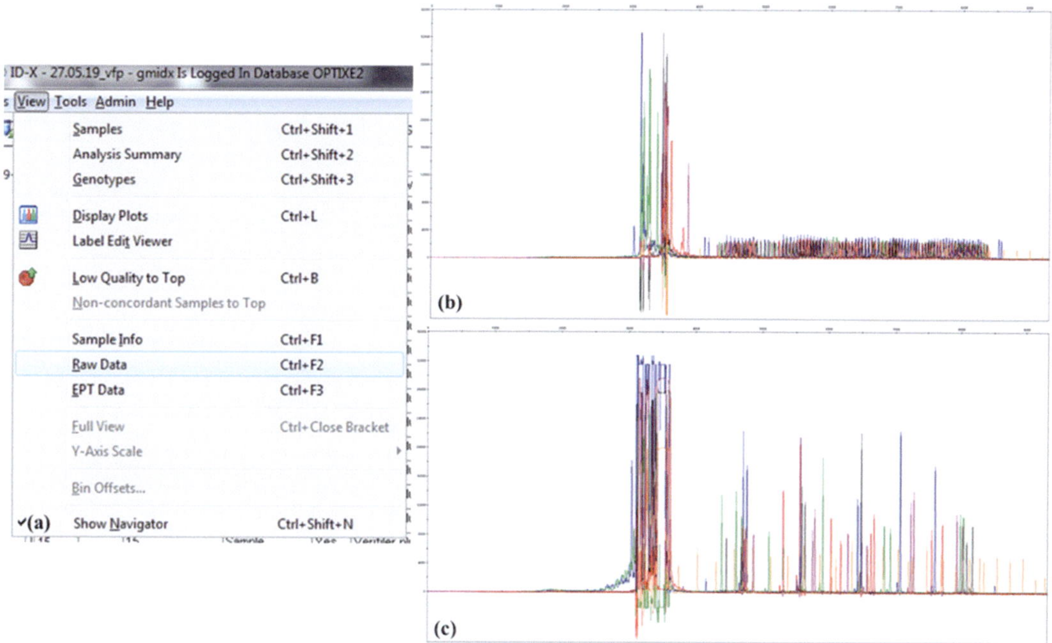

Fig. 24.17 Selection and visualization of (**a**) raw data for (**b**) allelic ladder and (**c**) sample

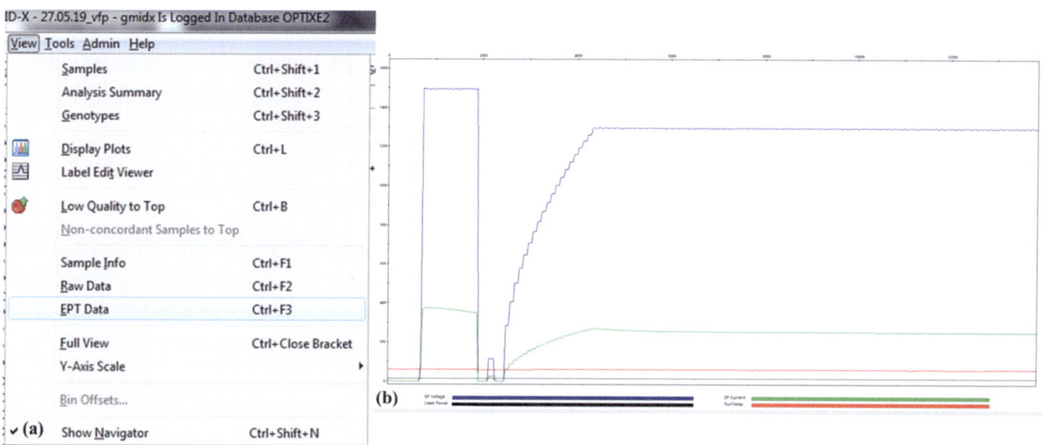

Fig. 24.18 (**a**) Visualization of "EPT data" during data analysis and (**b**) a representative EPT data

16. To check heterozygosity imbalance between two peaks, it is important to determine the peak height. Select "Tools" followed by "Plot Settings." Go to "Labels" and in "Label 2" select "Height," and press "OK." The peaks will be shown with allele number as well as peak height (Fig. 24.25).
17. Optionally you can modify other parameters of sample analysis such as "Peak detection threshold," "marker specific stutter ratio," "Peak quality parameters," and other "SQ and GQ parameters" by opening the analysis method, modifying the

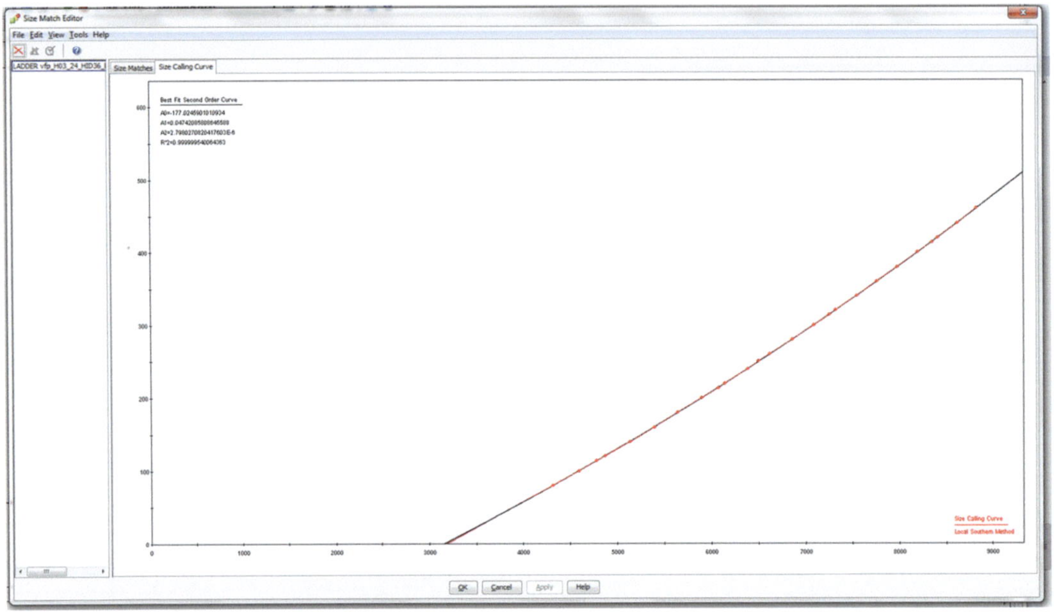

Fig. 24.19 Visualization of size calling curve

desired parameters, saving the analysis method, and reanalyzing the project (Fig. 24.26).

18. Select sizing table at the top right corner of the window after selecting a sample. Allele numbers of each marker amplified will appear as a separate window (Fig. 24.27).
19. Go to "File" and "Export Table" and save the allele table in the designated place in your system (Fig. 24.28).
20. After completion of analysis and transfer of allele data and other relevant information save the project prior to ending the GeneMapper® software (Fig. 24.29).

6 Observation

Under normal circumstances:

1. Autosomal STR profile of a single-source sample shows either two peaks for heterozygous alleles or one peak for homozygous allele.
2. For a normal sample the amelogenin marker shows either XY with balanced peak height for female sources, only X peak for female samples, or an unbalanced XY peak for mixed-source samples (not true for Y deletion samples).
3. Y chromosome STR profile of a single-source sample shows a single peak in all the markers due to its hemizygous nature and

Fig. 24.20 (**a**) Visualization of plots for allelic ladder and (**b**) a representative allelic ladder plot

more than one peaks appearing in mixed samples (in commercially available kits DYS385 and DYF387S1 markers show two peaks; however, these two markers are duplicated loci and considered as two independent loci).

4. X chromosome STR profile always shows a single peak for male single-source samples, whereas for single-source female samples it may show two peaks for heterozygous alleles or one peak for homozygous alleles.

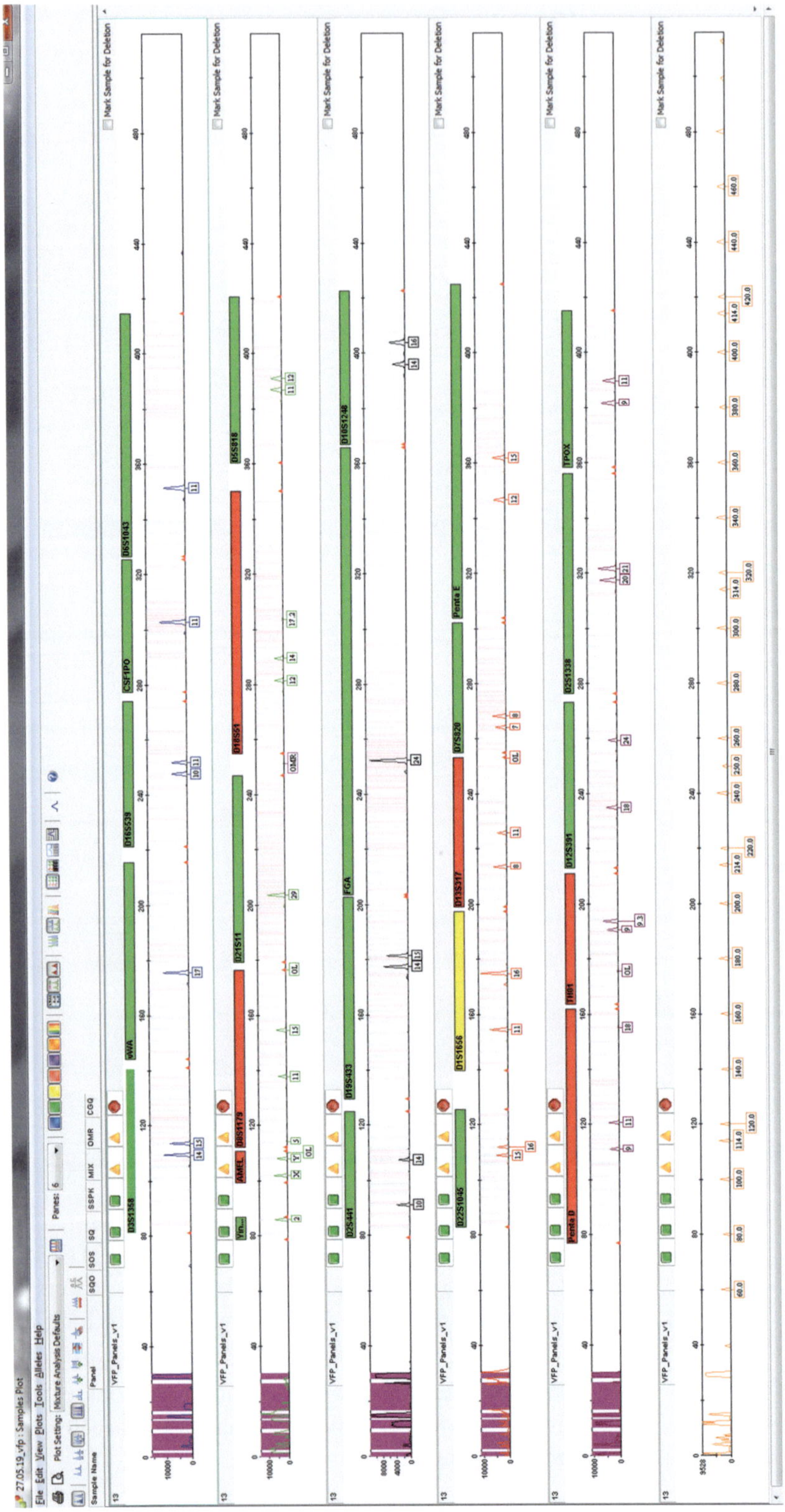

Fig. 24.21 Display of peaks in the markers analyzed in all six-dye sets

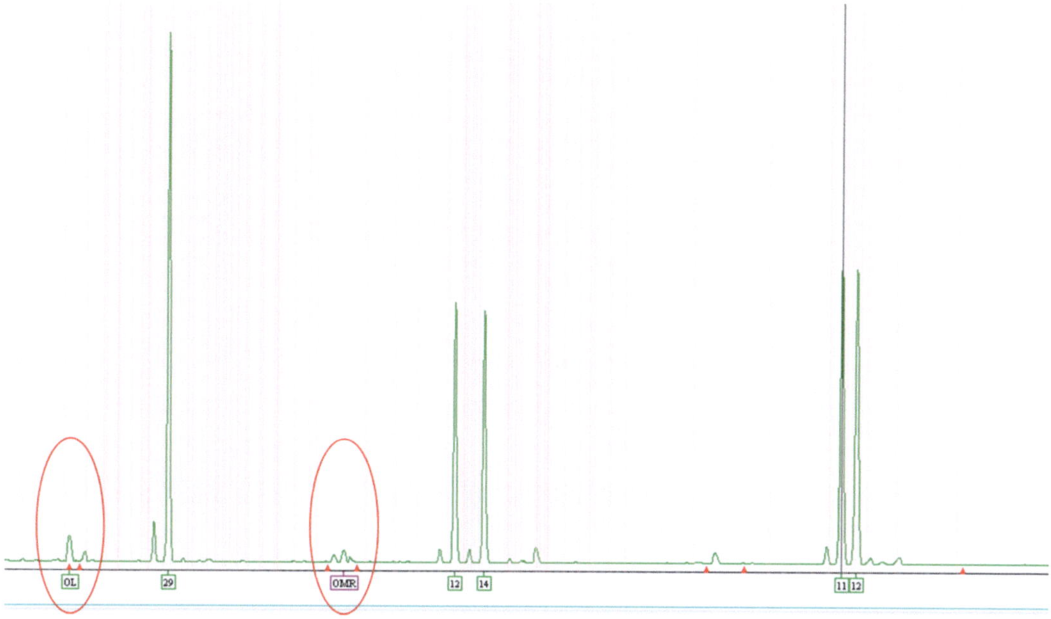

Fig. 24.22 Two common artifacts "OL" and "OMR" appearing in a profile

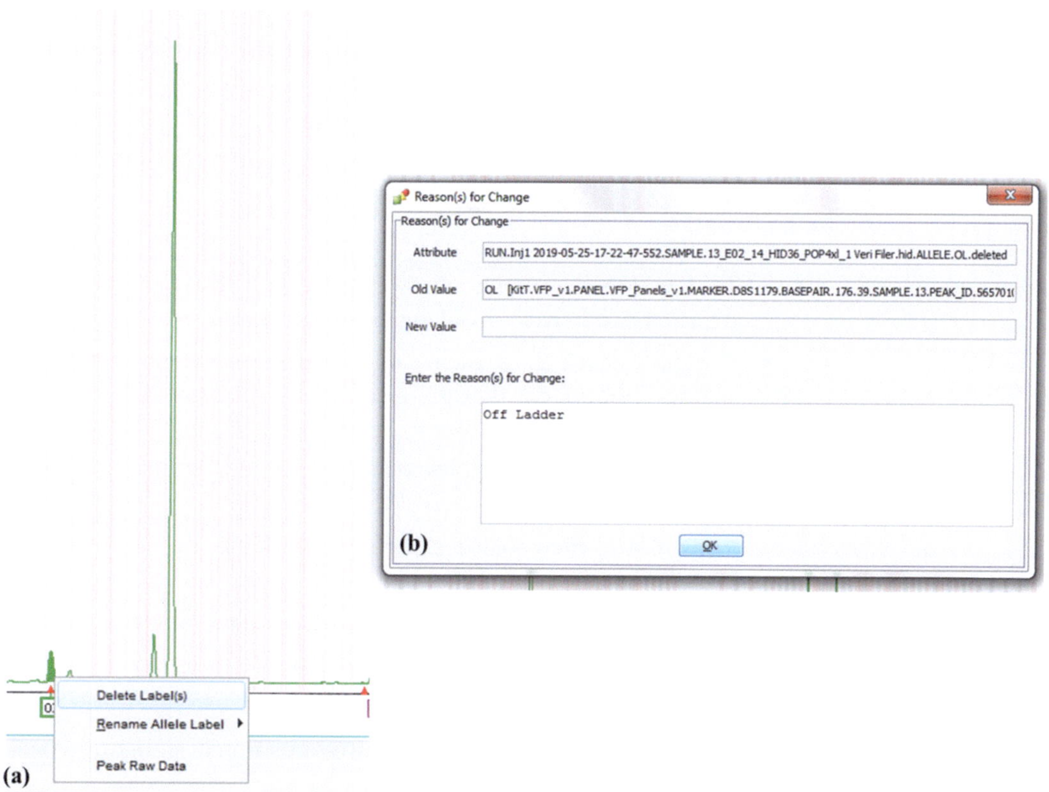

Fig. 24.23 (**a**) Correction of a common artifact "off-ladder" in the profile and (**b**) mentioning the reason for the change

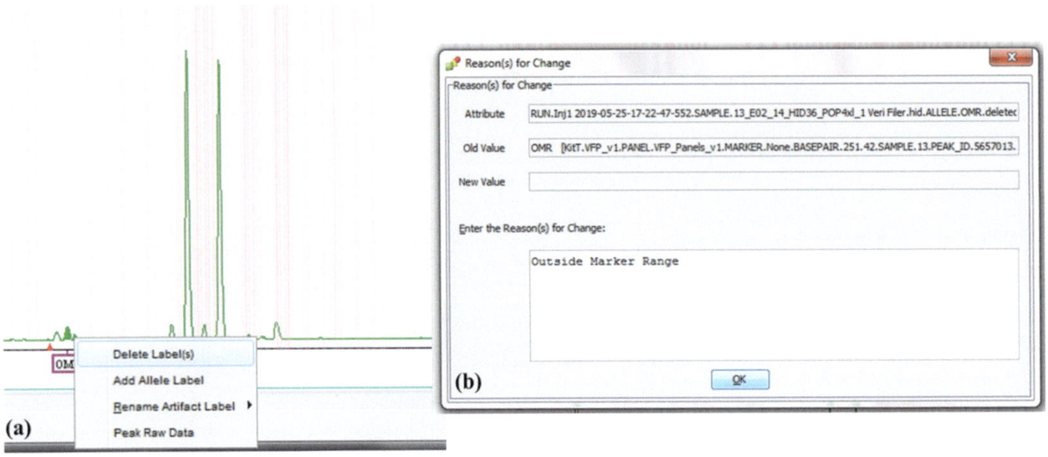

Fig. 24.24 (**a**) Correction of a common artifact "outside marker range" in the profile and (**b**) mentioning the reason for the change

7 Precautions

- Make sure to select the samples, allelic ladder, and controls correctly prior to sample analysis.
- Check size standard for the presence of any nonspecific peaks.
- Check ladder carefully as all the alleles should be called properly; none of the alleles should be called as OL.
- Never analyze any sample with less than 50 thresholds.
- Always check the inter-locus and intra-locus variation of peak height.
- In low DNA samples allele calls above 300 bps should be checked carefully to avoid any allele dropouts.
- Carefully check the presence of any additional allele beyond the marker boundary, though rarely this phenomenon can also occur.
- Analyze only those samples with all threshold parameters passed.
- Population-specific mutation rate of each marker should be known to the analyzer before working on any human population.
- Do not remove any labels with potential allele label.
- Extra caution should be adopted for mixed samples.
- Edit only the electrophoresis artifacts.
- If not masked, do not edit the stutter position designating a peak in mixed samples.
- Document all removal of labels.

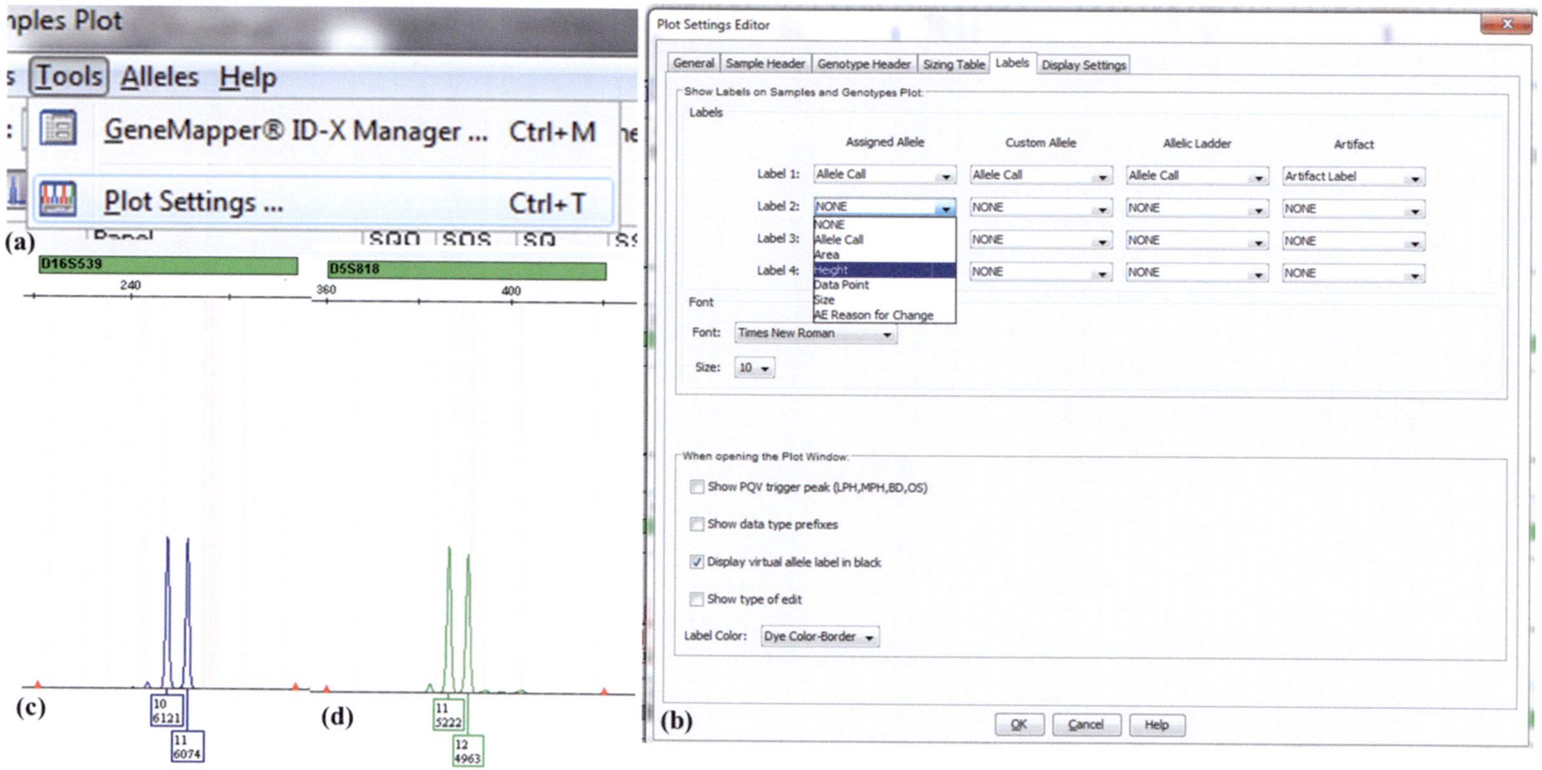

Fig. 24.25 Designation of both allele number and peak height from (**a**) Plot Settings (**b**) Plot Settings Editor in (**c, d**) a single electropherogram

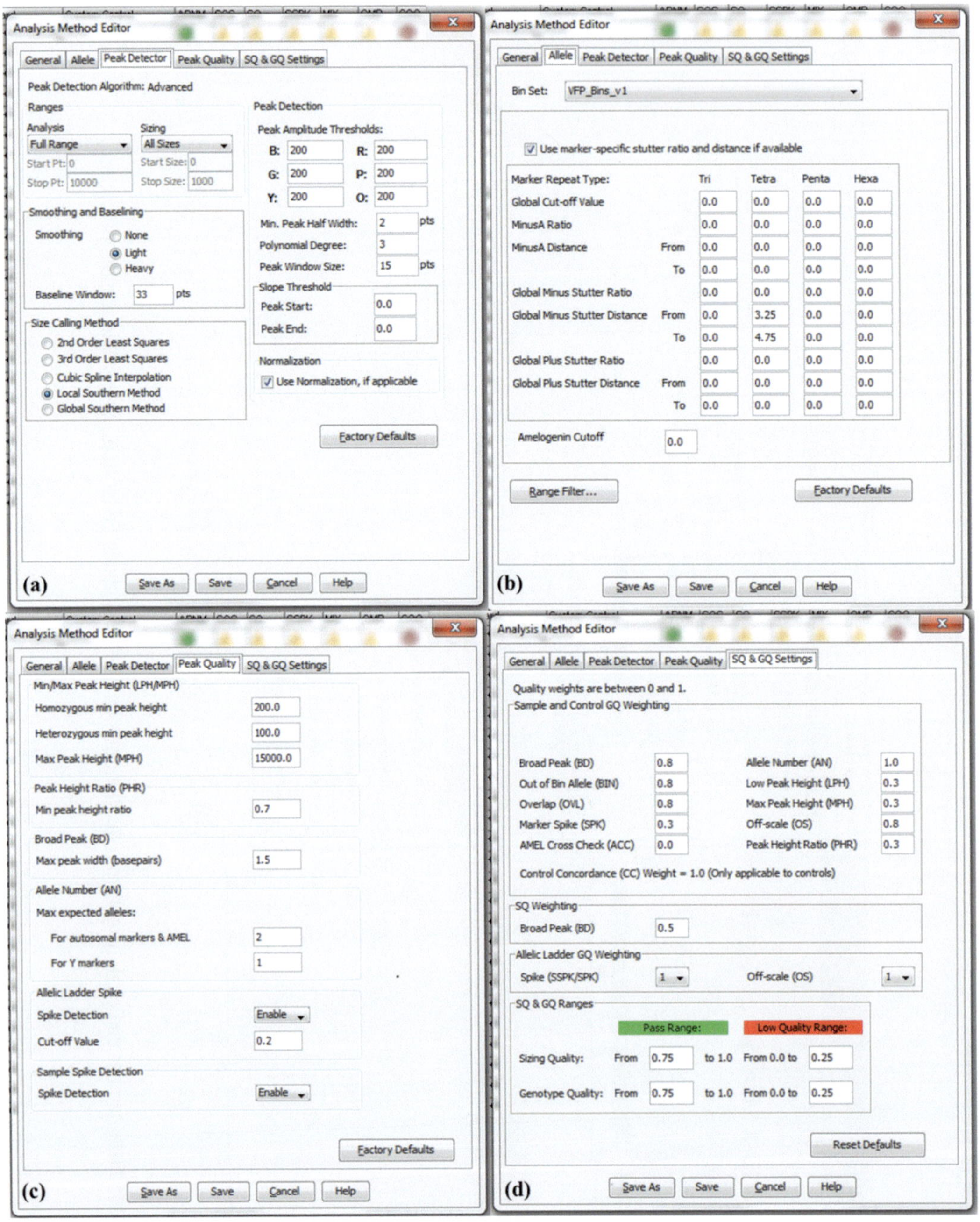

Fig. 24.26 Modification of various parameters of analysis such as (**a**) peak detection threshold, (**b**) minus and plus stutter distance, (**c**) peak quality parameters, and (**d**) SQ and GQ settings

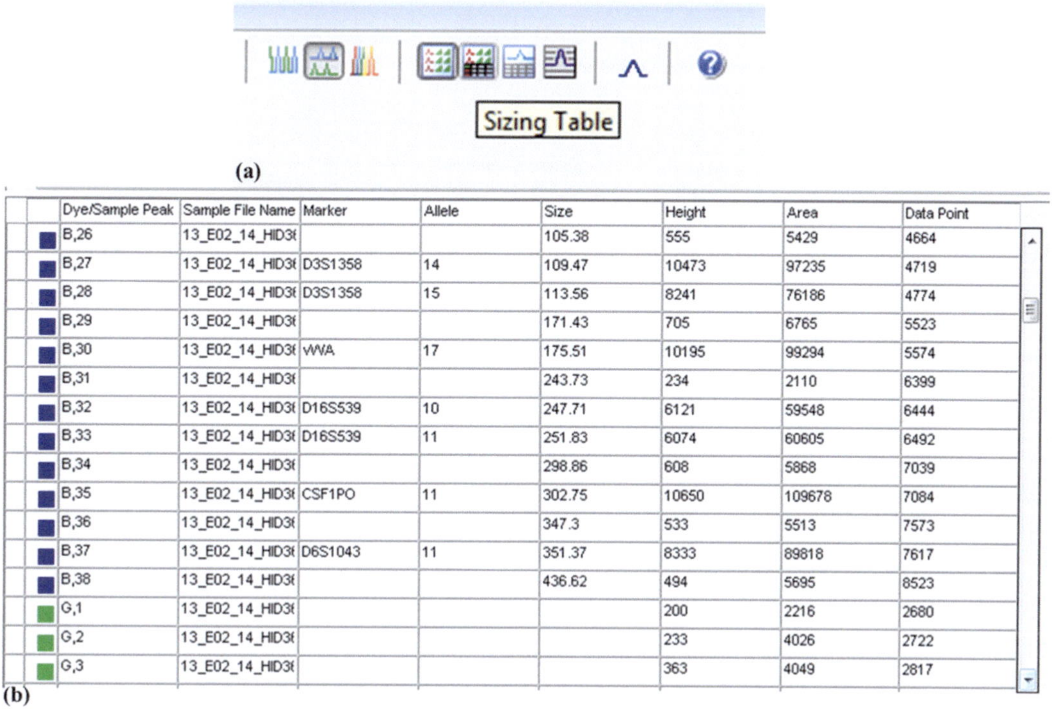

	Dye/Sample Peak	Sample File Name	Marker	Allele	Size	Height	Area	Data Point
	B,26	13_E02_14_HID3(			105.38	555	5429	4664
	B,27	13_E02_14_HID3(	D3S1358	14	109.47	10473	97235	4719
	B,28	13_E02_14_HID3(	D3S1358	15	113.56	8241	76186	4774
	B,29	13_E02_14_HID3(			171.43	705	6765	5523
	B,30	13_E02_14_HID3(	vWA	17	175.51	10195	99294	5574
	B,31	13_E02_14_HID3(			243.73	234	2110	6399
	B,32	13_E02_14_HID3(	D16S539	10	247.71	6121	59548	6444
	B,33	13_E02_14_HID3(	D16S539	11	251.83	6074	60605	6492
	B,34	13_E02_14_HID3(			298.86	608	5868	7039
	B,35	13_E02_14_HID3(	CSF1PO	11	302.75	10650	109678	7084
	B,36	13_E02_14_HID3(			347.3	533	5513	7573
	B,37	13_E02_14_HID3(	D6S1043	11	351.37	8333	89818	7617
	B,38	13_E02_14_HID3(			436.62	494	5695	8523
	G,1	13_E02_14_HID3(				200	2216	2680
	G,2	13_E02_14_HID3(				233	4026	2722
	G,3	13_E02_14_HID3(				363	4049	2817

Fig. 24.27 Window showing (**a**) sizing table of each marker along with (**b**) other parameters

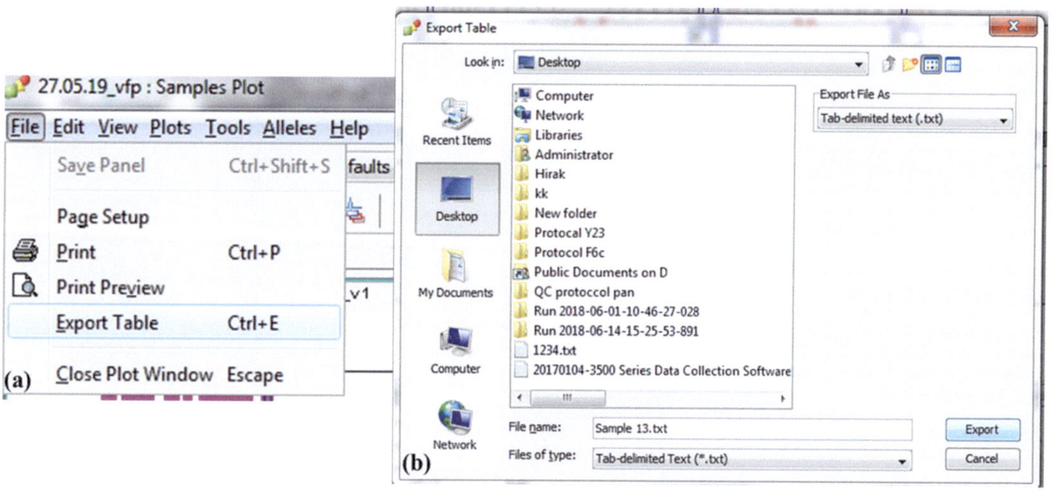

Fig. 24.28 (**a**, **b**) Window showing how to export sizing tables of the samples analyzed

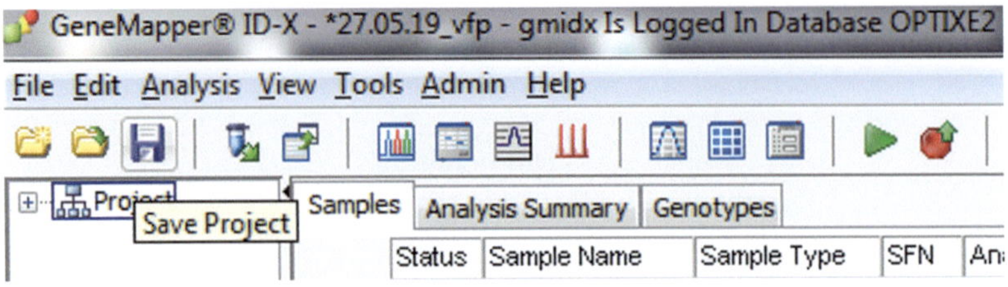

Fig. 24.29 Window showing how to save the project after completion of all the analyses and export of data

- Always check raw data, EPT, size calling curve, peaks of allelic ladder, size standard, and positive and negative control before analyzing any sample data.
- Regularly export your analyzed data to another file in the system.

8 Troubleshooting

Problem	Possible cause	Possible solutions
Quality failed (red)	Ladder failed	Check the raw data of ladder for the presence of peaks; if raw data shows the presence of high peaks, increase the threshold and reanalyze
		Check the raw data of ladder for the presence of peaks; if raw data shows the presence of low peaks, decrease the threshold and reanalyze
	Ladder not marked correctly	Assign the correct sample to be allelic ladder and reanalyze
	Peaks of ladder are marked as OL due to shift due to mobility issue in electrophoresis	Check the polymer and buffers; rerun with fresh reagents and repeat analysis
	Sample size too much	Dilute the amplicon and rerun
		Dilute the sample and re-PCR

(continued)

Problem	Possible cause	Possible solutions
Quality check (yellow)	Peaks do not fit to the bin sets	Mark allelic ladder properly and reanalyze
	Sample defined as positive control is not correct	Correctly define positive control and reanalyze
	Sample defined as negative control shows the presence of peaks	Rerun the samples with correct negative control
	Matrix not found	Locate the missing file or recreate it and reanalyze
	Too much size standard loaded	Decrease the quantity of size standard in subsequent runs
	Too much sample loaded	Decrease the quantity of samples in subsequent runs
	A difference in peak height in heterozygosity	Check the threshold and assign the correct alleles
No peaks detected	Amplification not occurred	Re-PCR the sample with optimum quantity of DNA
	Negative control marked as sample	Check the correct sample position
	High threshold maintained	Decrease the threshold to as low as 50 and reanalyze
Partial profile	Less DNA input	Use optimum quantity of DNA sample for PCR
	Degraded DNA sample	Use alternative STR kits such as mini-STRs or a kit containing mini-STR loci
	Presence of inhibitors	Modify the DNA extraction procedure and repeat the steps
	High threshold maintained	Decrease the threshold to as low as 50 and reanalyze

Chapter 25

Calculation of Paternity Index in Paternity Dispute and Identification Cases

Objective: To calculate the paternity index of an individual in cases of paternity dispute and identification.

1 Introduction

Cases related to paternity dispute of both criminal and civil natures can only be solved by DNA fingerprinting analysis. Genetic identification of a questioned individual to be the biological father/mother is the goal of this experiment. Besides paternity dispute this experiment also allows for reverse paternity testing when identification of any mutilated body or skeleton needs to be performed with the availability of parents' reference samples. Similar to paternity dispute cases, maternity dispute cases can also be solved using this technique where the father is known to be the biological father of the questioned child and the mother becomes alleged.

Half of a child's genetic material is inherited from its father and another half from its mother. The alleles of all the analyzed genetic markers of the disputed child are determined and compared with the alleles of both the parents followed by the application of a series of statistical equations to determine the probability of paternity. It is always a practice to calculate the paternity index when all the alleles of each genetic marker of the child match with the alleles of the parents. Though theoretically mismatch at a single locus is sufficient enough to exclude a person to be the biological father of the questioned child, mutational events cannot be ignored completely for one or few more allele mismatch. As currently we use more numbers of autosomal STR markers for paternity dispute cases, mutational events giving rise to mismatch at two genetic loci which occurs rarely can be considered to be a common event. In this regard, a number of software are available nowadays to

H. R. Dash et al., *Principles and Practices of DNA Analysis: A Laboratory Manual for Forensic DNA Typing*, Springer Protocols Handbooks, https://doi.org/10.1007/978-1-0716-0274-4_25, © Springer Science+Business Media, LLC, part of Springer Nature 2020

calculate the paternity index as well as kinship likelihood. Some of these computer tools include DNA-view (Charles Brenner, USA), Easy DNA (Wing Kam Fung, Hong Kong), EasyPat (Michael Krawczak, Germany), eDNA (eDNA Consortium), familias (Petter Mostad, Norway), FamLink (David Kling, Norway), GenoProof (Qualitype, Germany), GeneMarker HID (SoftGenetics, USA), and KinCalc (Steven Myers, USA). However, this chapter describes the manual calculation of paternity index in a disputed paternity case using MS Office Excel sheet.

2 Principle

The fundamental of rules of genetic inheritance of alleles in a generation involves the following:

(a) *Any genetic marker on autosomal chromosome of a child possesses two alleles, each of them inherited either from the father or the mother.*

(b) *Child always possesses haplotype of mitochondrial DNA from the matrilineal lineage.*

(c) *Only male child possesses Y chromosome haplotype from patrilineal lineage.*

In this regard, autosomal STRs are analyzed to calculate the paternity index (PI) which describes the probability of paternity of an individual. Calculation of PI is done as follows:

$$PI = X/Y$$

where

X = alleged father is the biological father

Y = any randomly selected man of that population is the biological father

This calculation gives rise to either inclusion or exclusion of the alleged father to be the biological father of the child with some statistical weight on the likelihood ratio in the form of paternity index (PI). In such cases, with the conventional matching of one-to-one allele on each marker, the obligate parental allele is determined and analyzed. As under normal circumstances the child receives one allele from each parent (Fig. 25.1), the probability of that allele present in a random man in the population is the basis of the statistical evaluation.

The must prerequisite of paternity index calculation involves the detailed population genetics study of the said population before any calculation. Allele frequency data of all the markers tested should be available for the population tested. On availability of DNA profiles from the alleged father, known mother and disputed

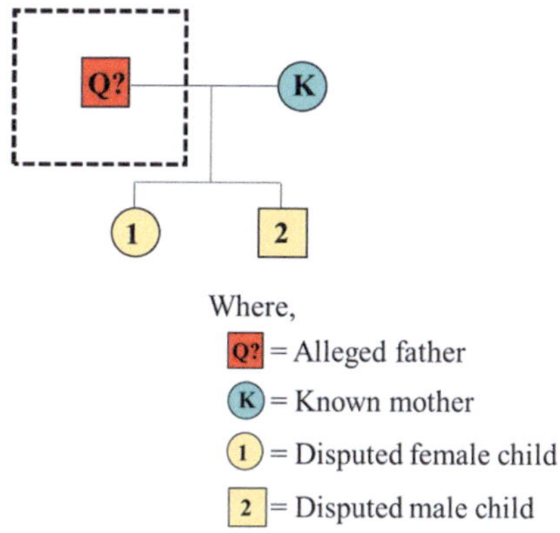

Fig. 25.1 Overview of a case of disputed paternity

child Punnett square is used to estimate the possibility of alleged parents' genotypes to produce the child's genotype. Further allele frequency data of each genetic marker comes into play to exclude the false inclusion of any random men in this case.

As current-day DNA fingerprinting technique involves the typing of multiple genetic loci simultaneously, it increases the discrimination power. In such cases individual paternity index (PI) is calculated for all each locus separately and the product of all the individual PIs gives rise to combined paternity index (CPI). CPI signifies the chance of any random, unrelated man of the population to be the biological father. Using this CPI value the probability of paternity of the alleged individual can be calculated using the formula:

$$\text{Probability of paternity} = (\text{CPI}/\text{CPI} + 1) \times 100$$

3 Requirements

3.1 Autosomal STR DNA Profiles

For a paternity trio experiment to be performed the autosomal STR DNA profile is required for the questioned child, known mother, and alleged father. Genetic markers should be amplified using a single multiplex kit for all three samples. Additionally, all three samples need to be run in a single electrophoresis experiment. The parameters set for analyzing the obtained profiles should be the same for these three samples. For analysis in this experiment, the profile obtained for the paternity trio experiment is as follows:

Genetic markers	Known mother	Male child	Alleged father
D3S1358	16,17	16,17	16,17
vWA	14,17	17,19	18,19
D16S539	11,12	11,12	11,13

(continued)

Genetic markers	Known mother	Male child	Alleged father
CSF1PO	11,12	10,12	10,12
TPOX	11,11	8,11	8,9
D8S1179	10,10	10,16	16,16
D21S11	28,33.2	31,33.2	30,31
D18S51	13,20	14,20	14,18
D2S441	11,13	11,13	10,11
D19S433	13,14	13,13.2	13.2,14
TH01	6,9	6,8	7,8
FGA	20,23	22,23	22,22
D22S1045	15,16	15,15	15,16
D5S818	10,11	10,11	11,13
D13S317	8,10	10,12	12,12
D7S820	8,8	8,8	8,8
SE33	19,28.2	28.2,30.2	30.2,31.2
D10S1248	14,17	14,17	14,17
D1S1656	11,11	11,12	12,17
D12S391	19,21	17,19	17,18.3
D2S1338	20,24	20,22	19,22
Amelogenin	XX	XY	XY
DYS391	–	11	11
InDel	–	2	2

3.2 Allele Frequencies

Allele frequency data is required for performing the calculation. It describes the proportion of total number of alleles represented by a particular allele in a population. This is population specific. Due to polymorphic nature the STR alleles may generate many numbers of alleles in a population. For the total allele pool in a population, homozygous individuals contribute two of the same allele whereas heterozygous individuals contribute one of the particular alleles to the total allele number of the population; for example, in a population of 100 individuals, 15 individuals are homozygous for allele M (MM), 40 individuals are heterozygous MN, and rest 45 individuals are heterozygous MO. Thus, 15 homozygous (MM) individuals contribute two "M" allele each in the population, 40 heterozygous (MN) individuals contribute one "M" allele each to the population whereas another 45 heterozygous (MO) individuals also contribute one "M" allele each to the population.

Thus, in a three-allele system population, the allele frequency can be calculated as

$$\text{Frequency of allele}\,''\text{M}'' = [(2 \times \text{No. of homozygous MM individuals}) + (\text{No. of heterozygous MN individuals}) + (\text{No. of homozygous MO individuals})]/2n$$

where n = no. of individuals in the populations.

The sum total of the frequency of all the alleles in a population for a genetic marker will be one. Allele frequency data calculated earlier for central Indian population is given in Appendix D which will be used in this experiment.

3.3 MS Office Excel Platform

All the calculations will be performed in MS Office Excel sheet. Hence, it should be installed in the system prior to the commencement of the PI calculation.

4 Procedure

1. Open a new Excel sheet using MS Office tool (Fig. 25.2).
2. Enter genetic markers and their corresponding genotype for the known mother and the questioned child side by side (Fig. 25.3).
3. Calculate "obligate allele" for each marker. Obligate allele is either two or one in number (Fig. 25.4). It is the allele that must be present in the alleged father if he is the correct biological father. In the current example, for genetic marker "D3S1358" both 16 and 17 can be the obligate alleles, as both 16 and 17 have the possibility of being transmitted from the known mother. However, for "vWA" 19 becomes the obligate allele as it is evidenced that 17 is inherited from the mother to the questioned child and 19 must be present in the alleged father to be the biological father. Significance of the

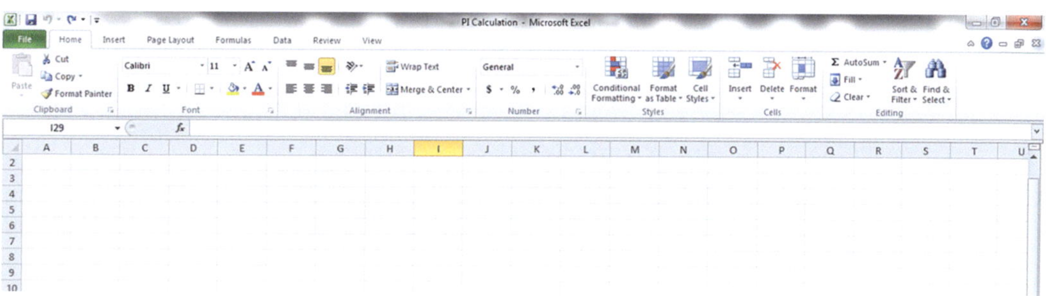

Fig. 25.2 A blank MS Office Excel sheet

D	E	F
Genetic Markers	**Known Mother**	**Questioned Child**
D3S1358	16,17	16,17
vWA	14,17	17,19
D16S539	11,12	11,12
CSF1PO	11,12	10,12
TPOX	11,11	8,11
D8S1179	10,10	10,16
D21S11	28,33.2	31,33.2
D18S51	13,20	14,20
D2S441	11,13	11,13
D19S433	13,14	13,13.2
TH01	6,9	6,8
FGA	20,23	22,23
D22S1045	15,16	15,15
D5S818	10,11	10,11
D13S317	8,10	10,12
D7S820	8,8	8,8
SE33	19,28.2	28.2,30.2
D10S1248	14,17	14,17
D1S1656	11,11	11,12
D12S391	19,21	17,19
D2S1338	20,24	20,22

Fig. 25.3 Genetic profile of the known mother and questioned child

"obligate allele" is that when the obligate paternal allele has corresponding allele in the alleged father, it is a case of inclusion. In contrary, when the obligate paternal allele does not have the corresponding allele in the alleged father, it becomes a case of exclusion.

4. Enter the profile of the alleged father besides the obligate allele (Fig. 25.5). Check whether the obligate allele is present in the corresponding genetic marker of the alleged father or not.
5. Enter allele frequency of each obligate allele next to the profile of the alleged father (Fig. 25.6). As allelic frequency data is population specific, it should be calculated prior to performance of the experiment or the data from any published literature can be used for this purpose. In this experiment, we will use the allele frequency data calculated on central Indian population which is given in Appendix D.
6. Calculate the likelihood that the alleged father could transmit the obligate allele "X" (Fig. 25.7). The value of "X" is either 1 or 0.5. When the father is homozygous (see FGA of the current example) the value of X becomes 1. Similarly, when both the obligate alleles are present in the alleged father's genotype (see D3S1358 of the current example) the total value of "X" becomes 1. Thus, the value of "X" becomes 1 as

D	E	F	G
Genetic Markers	Known Mother	Questioned Child	Obligate Allele
D3S1358	16,17	16,17	16
			17
vWA	14,17	17,19	19
D16S539	11,12	11,12	11
			12
CSF1PO	11,12	10,12	10
TPOX	11,11	8,11	8
D8S1179	10,10	10,16	16
D21S11	28,33.2	31,33.2	31
D18S51	13,20	14,20	14
D2S441	11,13	11,13	11
			13
D19S433	13,14	13,13.2	13.2
TH01	6,9	6,8	8
FGA	20,23	22,23	22
D22S1045	15,16	15,15	15
D5S818	10,11	10,11	10
			11
D13S317	8,10	10,12	12
D7S820	8,8	8,8	8
SE33	19,28.2	28.2,30.2	30.2
D10S1248	14,17	14,17	14
			17
D1S1656	11,11	11,12	12
D12S391	19,21	17,19	17
D2S1338	20,24	20,22	22

Fig. 25.4 Determination of obligate paternal allele from the profile of known mother and questioned child

the likelihood of transmitting the allele(s) becomes $2/2 = 1$. In contrary, when any one of the two obligate alleles is present in the alleged father's genotype (see D16S539 of the current example) or the alleged father has only one copy of the obligate allele (see CSF1PO of the current example) the value of X becomes 0.5. This is due to the fact that the likelihood of transmitting the obligate allele from the paternal genotype to the disputed child becomes $1/2 = 0.5$.

7. Calculate paternity index (PI) using the formula X/allele frequency, where X is the likelihood of transmitting the paternal obligate allele by the alleged father. Thus, PI will be either 1/allele frequency or 0.5/allele frequency. However, in certain instances (as in case of D3S1358 and D16S539) there are two

D	E	F	G	H
Genetic Marker:	Known Mother	Questioned Chilc	Obligate Allele	Alleged Father
D3S1358	16,17	16,17	16 17	16,17
vWA	14,17	17,19	19	18,19
D16S539	11,12	11,12	11 12	11,13
CSF1PO	11,12	10,12	10	10,12
TPOX	11,11	8,11	8	8,9
D8S1179	10,10	10,16	16	16,16
D21S11	28,33.2	31,33.2	31	30,31
D18S51	13,20	14,20	14	14,18
D2S441	11,13	11,13	11 13	10,11
D19S433	13,14	13,13.2	13.2	13.2,14
TH01	6,9	6,8	8	7,8
FGA	20,23	22,23	22	22,22
D22S1045	15,16	15,15	15	15,16
D5S818	10,11	10,11	10 11	11,13
D13S317	8,10	10,12	12	12,12
D7S820	8,8	8,8	8	8,8
SE33	19,28.2	28.2,30.2	30.2	30.2,31.2
D10S1248	14,17	14,17	14 17	14,17
D1S1656	11,11	11,12	12	12,17
D12S391	19,21	17,19	17	17,18.3
D2S1338	20,24	20,22	22	19,22

Fig. 25.5 Genetic profile of the alleged father besides obligate alleles

obligate alleles and hence two corresponding allele frequencies. In this case, the PI is calculated as 1/(allele frequency of 16 + allele frequency of 17) (for D3S1358) or 0.5/(allele frequency of 11 + allele frequency of 12) (for D16S539) (Fig. 25.8).

8. Calculate combined paternity index (CPI). It can be calculated by multiplication of the individual PIs of the individual genetic markers. In this case the CPI value is calculated to be 5.25576×10^{11} (Fig. 25.9).
9. Calculate random man not excluded (RMNE) value. It signifies the power of a paternity test by determining the proportion of the population that are capable of contributing to the obligate allele and hence cannot be excluded or included falsely. It can be calculated by using the formula RMNE $= 1 - (1 - \text{frequency})^2$ or $(\text{frequency})^2 + 2\ (\text{frequency})\ (1 - \text{frequency})$ (Fig. 25.10).

D	E	F	G	H	I
Genetic Marker:	**Known Mother**	**Questioned Chilc**	**Obligate Allele**	**Alleged Father**	**Allele Frequency**
D3S1358	16,17	16,17	16	16,17	0.315
			17		0.252
vWA	14,17	17,19	19	18,19	0.076
D16S539	11,12	11,12	11	11,13	0.29
			12		0.235
CSF1PO	11,12	10,12	10	10,12	0.188
TPOX	11,11	8,11	8	8,9	0.339
D8S1179	10,10	10,16	16	16,16	0.106
D21S11	28,33.2	31,33.2	31	30,31	0.035
D18S51	13,20	14,20	14	14,18	0.293
D2S441	11,13	11,13	11	10,11	0.36
			13		0.02
D19S433	13,14	13,13.2	13.2	13.2,14	0.021
TH01	6,9	6,8	8	7,8	0.142
FGA	20,23	22,23	22	22,22	0.159
D22S1045	15,16	15,15	15	15,16	0.364
D5S818	10,11	10,11	10	11,13	0.108
			11		0.389
D13S317	8,10	10,12	12	12,12	0.274
D7S820	8,8	8,8	8	8,8	0.219
SE33	19,28.2	28.2,30.2	30.2	30.2,31.2	0.075
D10S1248	14,17	14,17	14	14,17	0.262
			17		0.085
D1S1656	11,11	11,12	12	12,17	0.098
D12S391	19,21	17,19	17	17,18.3	0.135
D2S1338	20,24	20,22	22	19,22	0.055

Fig. 25.6 Window showing allele frequency data of the corresponding obligate alleles

When more than two allele frequencies are available for the same locus (as in case of D3S1358 and D16S539 loci), the calculation should be performed using the formula RMNE = 1 − [1 − (frequency 1 + frequency 2)^2]. Additionally calculate the combined random men not excluded (CRMNE) value (Fig. 25.10). It can be calculated by simply multiplying the individual calculated RMNE values. Due to very small values of CRMNE, it is recommended to calculate the reciprocal of this number (1/CRMNE), which is referred to as "probability of exclusion" or exclusion power (EP) or power of exclusion (PE) (Fig. 25.10). It represents the probability of excluding a falsely accused man. In this case, the calculated value of PE is 56,171,746,392. It signifies that the alleged father is approximately 56,171,746,392 times as likely to be the father of the questioned child as an unrelated Indian male.

D	E	F	G	H	I	J
Genetic Markers	**Known Mother**	**Questioned Child**	**Obligate Allele**	**Alleged Father**	**Allele Frequency**	**X**
D3S1358	16,17	16,17	16	16,17	0.315	1
			17		0.252	
vWA	14,17	17,19	19	18,19	0.076	0.5
D16S539	11,12	11,12	11	11,13	0.29	0.5
			12		0.235	
CSF1PO	11,12	10,12	10	10,12	0.188	0.5
TPOX	11,11	8,11	8	8,9	0.339	0.5
D8S1179	10,10	10,16	16	16,16	0.106	1
D21S11	28,33.2	31,33.2	31	30,31	0.035	0.5
D18S51	13,20	14,20	14	14,18	0.293	0.5
D2S441	11,13	11,13	11	10,11	0.36	0.5
			13		0.02	
D19S433	13,14	13,13.2	13.2	13.2,14	0.021	0.5
TH01	6,9	6,8	8	7,8	0.142	0.5
FGA	20,23	22,23	22	22,22	0.159	1
D22S1045	15,16	15,15	15	15,16	0.364	0.5
D5S818	10,11	10,11	10	11,13	0.108	
			11		0.389	0.5
D13S317	8,10	10,12	12	12,12	0.274	1
D7S820	8,8	8,8	8	8,8	0.219	1
SE33	19,28.2	28.2,30.2	30.2	30.2,31.2	0.075	0.5
D10S1248	14,17	14,17	14	14,17	0.262	1
			17		0.085	
D1S1656	11,11	11,12	12	12,17	0.098	0.5
D12S391	19,21	17,19	17	17,18.3	0.135	0.5
D2S1338	20,24	20,22	22	19,22	0.055	0.5

Fig. 25.7 Calculation of likelihood of transmitting paternal obligate allele from the alleged father

10. Calculate the probability of paternity (PP). PP reflects the probability of the alleged father to be the biological father of the questioned child. It can be calculated using the formula: $PP = 1/1 + (1/CPI)$. It has been calculated to be 0.999999999998 for the current analysis.

5 Observation

The probability of paternity in this case is calculated to be 0.999999999998 which is >99.9999%. This suggests that the alleged father cannot be excluded as the biological father of the questioned child, under the assumption that the combined paternity index is 5.25576×10^{11}, which results in the probability of paternity of >99.9999% when compared to an untested, unrelated individual of the Indian population (assuming the prior probability of 0.5).

D	E	F	G	H	I	J	K
Genetic Markers	Known Mother	Questioned Child	Obligate Allele	Alleged Father	Allele Frequency	X	PI
D3S1358	16,17	16,17	16	16,17	0.315	1	1.7636
			17		0.252		
vWA	14,17	17,19	19	18,19	0.076	0.5	6.57895
D16S539	11,12	11,12	11	11,13	0.29	0.5	0.9523
			12		0.235		
CSF1PO	11,12	10,12	10	10,12	0.188	0.5	2.65957
TPOX	11,11	8,11	8	8,9	0.339	0.5	1.47493
D8S1179	10,10	10,16	16	16,16	0.106	1	9.43396
D21S11	28,33.2	31,33.2	31	30,31	0.035	0.5	14.2857
D18S51	13,20	14,20	14	14,18	0.293	0.5	1.70648
D2S441	11,13	11,13	11	10,11	0.36	0.5	1.3157
			13		0.02		
D19S433	13,14	13,13.2	13.2	13.2,14	0.021	0.5	23.8095
TH01	6,9	6,8	8	7,8	0.142	0.5	3.52113
FGA	20,23	22,23	22	22,22	0.159	1	6.28931
D22S1045	15,16	15,15	15	15,16	0.364	0.5	1.37363
D5S818	10,11	10,11	10	11,13	0.108		
			11		0.389	0.5	1.006
D13S317	8,10	10,12	12	12,12	0.274	1	3.64964
D7S820	8,8	8,8	8	8,8	0.219	1	4.56621
SE33	19,28.2	28.2,30.2	30.2	30.2,31.2	0.075	0.5	6.66667
D10S1248	14,17	14,17	14	14,17	0.262	1	2.8818
			17		0.085		
D1S1656	11,11	11,12	12	12,17	0.098	0.5	5.10204
D12S391	19,21	17,19	17	17,18.3	0.135	0.5	3.7037
D2S1338	20,24	20,22	22	19,22	0.055	0.5	9.09091

Fig. 25.8 Calculation of paternity index (PI) of the individual genetic markers

6 Precautions

- Perform all the calculations carefully.
- Always use population-specific allele frequency data for the calculation.
- Carefully determine the paternal obligate allele.
- Determine the likelihood value of transmitting the paternal obligate allele to be either 0.5 or 1.
- While calculating CPI or CRMNE values do not sum the individual PI or RMNE values; the correct formula is to multiply the individual values.
- The calculated CRMNE value is generally very small; be careful while calculating to obtain a smaller CRMNE value after calculation.
- For a case of inclusion the calculated probability of paternity value should be >0.99999.

D	E	F	G	H	I	J	K	L
Genetic Markers	**Known Mother**	**Questioned Child**	**Obligate Allele**	**Alleged Father**	**Allele Frequency**	**X**	**PI**	**CPI**
D3S1358	16,17	16,17	16	16,17	0.315	1	1.7636	5.25576E+11
			17		0.252			
vWA	14,17	17,19	19	18,19	0.076	0.5	6.57895	
D16S539	11,12	11,12	11	11,13	0.29	0.5	0.9523	
			12		0.235			
CSF1PO	11,12	10,12	10	10,12	0.188	0.5	2.65957	
TPOX	11,11	8,11	8	8,9	0.339	0.5	1.47493	
D8S1179	10,10	10,16	16	16,16	0.106	1	9.43396	
D21S11	28,33.2	31,33.2	31	30,31	0.035	0.5	14.2857	
D18S51	13,20	14,20	14	14,18	0.293	0.5	1.70648	
D2S441	11,13	11,13	11	10,11	0.36	0.5	1.3157	
			13		0.02			
D19S433	13,14	13,13.2	13.2	13.2,14	0.021	0.5	23.8095	
TH01	6,9	6,8	8	7,8	0.142	0.5	3.52113	
FGA	20,23	22,23	22	22,22	0.159	1	6.28931	
D22S1045	15,16	15,15	15	15,16	0.364	0.5	1.37363	
D5S818	10,11	10,11	10	11,13	0.108			
			11		0.389	0.5	1.006	
D13S317	8,10	10,12	12	12,12	0.274	1	3.64964	
D7S820	8,8	8,8	8	8,8	0.219	1	4.56621	
SE33	19,28.2	28.2,30.2	30.2	30.2,31.2	0.075	0.5	6.66667	
D10S1248	14,17	14,17	14	14,17	0.262	1	2.8818	
			17		0.085			
D1S1656	11,11	11,12	12	12,17	0.098	0.5	5.10204	
D12S391	19,21	17,19	17	17,18.3	0.135	0.5	3.7037	
D2S1338	20,24	20,22	22	19,22	0.055	0.5	9.09091	

Fig. 25.9 Spreadsheet showing the calculation of combined paternity index (CPI) value

D	E	F	G	H	I	J	K	L	M	N	O
Genetic Markers	**Known Mother**	**Questioned Child**	**Obligate Allele**	**Alleged Father**	**Allele Frequency**	**X**	**PI**	**CPI**	**RMNE**	**CRMNE**	**PE**
D3S1358	16,17	16,17	16	16,17	0.315	1	1.7636	5.25576E+11	0.75449	1.78025E-11	56171746392
			17		0.252						
vWA	14,17	17,19	19	18,19	0.076	0.5	6.57895		0.14622		
D16S539	11,12	11,12	11	11,13	0.29	0.5	0.9523		0.77438		
			12		0.235						
CSF1PO	11,12	10,12	10	10,12	0.188	0.5	2.65957		0.34066		
TPOX	11,11	8,11	8	8,9	0.339	0.5	1.47493		0.56308		
D8S1179	10,10	10,16	16	16,16	0.106	1	9.43396		0.20076		
D21S11	28,33.2	31,33.2	31	30,31	0.035	0.5	14.2857		0.06878		
D18S51	13,20	14,20	14	14,18	0.293	0.5	1.70648		0.50015		
D2S441	11,13	11,13	11	10,11	0.36	0.5	1.3157		0.6156		
			13		0.02						
D19S433	13,14	13,13.2	13.2	13.2,14	0.021	0.5	23.8095		0.04156		
TH01	6,9	6,8	8	7,8	0.142	0.5	3.52113		0.26384		
FGA	20,23	22,23	22	22,22	0.159	1	6.28931		0.29272		
D22S1045	15,16	15,15	15	15,16	0.364	0.5	1.37363		0.5955		
D5S818	10,11	10,11	10	11,13	0.108						
			11		0.389	0.5	1.006				
D13S317	8,10	10,12	12	12,12	0.274	1	3.64964		0.47292		
D7S820	8,8	8,8	8	8,8	0.219	1	4.56621		0.39004		
SE33	19,28.2	28.2,30.2	30.2	30.2,31.2	0.075	0.5	6.66667		0.14438		
D10S1248	14,17	14,17	14	14,17	0.262	1	2.8818		0.74699		
			17		0.085						
D1S1656	11,11	11,12	12	12,17	0.098	0.5	5.10204		0.1864		
D12S391	19,21	17,19	17	17,18.3	0.135	0.5	3.7037		0.25178		
D2S1338	20,24	20,22	22	19,22	0.055	0.5	9.09091		0.10698		

Fig. 25.10 Calculation of RMNE, CRMNE, and PE values

7 Troubleshooting

Problem	Possible cause	Possible solutions
Calculated probability of paternity is not >99.9999	Case of exclusion	Check the presence of all paternal obligate alleles in alleged father's profile
		Check all the calculations carefully before declaring the case of exclusion
	Calculation is not correct	Check all the calculations and formula used carefully
	Proper allele frequency data is not selected	Use population-specific allele frequency data
Excel sheet showing the probability of paternity value to be 1	Decimal places are selected by default	Go to Format Cells: Numbers: Select decimal places and increase the decimal places
Paternity trio not complete	Known mother profile not available	You cannot use the same calculations for this case
	Mother is also alleged	Draw the Punnett square to see whether the questioned child inherits alleles from both the parents or not
		First keep the mother's genotypes to be known and calculate paternity index using the same calculation
		Next keep the father's genotypes to be known and calculate maternity index using the same calculation

Part VII

Case Studies

Chapter 26

Solving Paternity Dispute by DNA Fingerprinting: A Case Study

Objective: To solve a paternity dispute case by DNA fingerprinting.

1 Introduction

Paternity dispute cases are of either criminal or civil in nature. Sexual assault in pretext of marriage leading to pregnancy can be solved by identification of fetus or newborn as the biological offspring of the victim and the suspect. Similarly, paternity dispute in civil nature is also common where the biological father denies the claim of the child. In both the scenario DNA fingerprinting is the only established technique to draw a conclusive decision to ascertain that the claims and the biological identity of an individual are established.

In such cases the reference biological samples of the questioned child, known mother, and alleged father are collected. The reference sample may include the invasive sample blood in EDTA vial or on FTA® card or the noninvasive sample such as buccal swab. Each laboratory has its guidelines for collection, preservation, and transportation of reference samples to the laboratory for examination. However, in this case maintenance of chain of custody is highly important. Liquid blood sample should be collected aseptically from the donor in EDTA vial and stored in ice till arrival in the laboratory within a stipulated period of time. Similarly, liquid blood can be spread on the designated area of the FTA® card, dried completely, and stored and transported at room temperature to the laboratory. Noninvasive samples such as buccal swab can also be collected and the cotton swab should be dried completely before its storage and transportation at room temperature. Additionally, buccal swab can also be prepared on a FTA® card and processed similarly to the blood sample on FTA® card.

H. R. Dash et al., *Principles and Practices of DNA Analysis: A Laboratory Manual for Forensic DNA Typing*, Springer Protocols Handbooks, https://doi.org/10.1007/978-1-0716-0274-4_26,

2 Principle

Autosomal STR analysis is conducted for solving paternity disputes cases of both civil and criminal nature. The nature of transfer of parental genotypes to the offspring is dependent on the allele combination present in the parents. The segregation of alleles in a genetic locus always follows the laws of Mendelian inheritance. For parentage examination the Mendelian rules should be tested as follows:

(a) The allele absent in both the parents cannot be present in the child.

(b) The pair of alleles is inherited from the parents.

(c) When both the parents have the same allele, there is the possibility that the child may possess two copies of the same allele.

(d) The genetic marker present as an identical pair in both the parents is present in the child.

Thus, in the absence of mutational events, establishment of parentage depends on the fact that on each genetic locus tested the alleles of the child match with the corresponding alleles of the parents at the same genetic marker (Fig. 26.1).

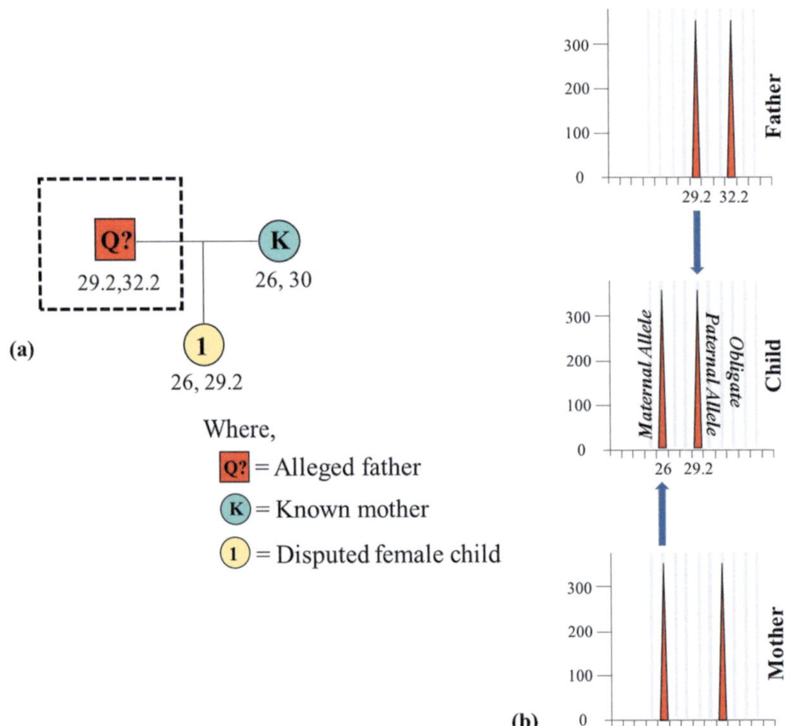

Fig. 26.1 Inheritance of alleles from both the parents to the offspring as per (**a**) Mendelian pattern of inheritance and (**b**) determination of maternal and obligate paternal allele

For solving paternity dispute cases, the ISFG Paternity Testing Commission (PTC) has recommended several considerations such as calculation of likelihood ratio (LR) of the DNA evidences and use of population-specific genetics for calculation of allele frequency as well as for cases with sample deficiency, extra statistical evaluations should be carried out before deducing any conclusion.

3 Brief Case History

In the pretext of marriage, the suspect performed sexual activity with the complainant. When the complainant became pregnant, the suspect denied of marrying her. A criminal case of sexual assault was filed in an Indian police station by the complainant. When the child was born from the complainant, the case was referred for DNA examination. Subsequently, the DNA fingerprinting laboratory received liquid blood samples of the complainant, newborn, and the suspect in EDTA vials through proper channel which were subjected to examination.

4 Samples

Three liquid blood samples in EDTA vial consisting of Article A (blood sample of complainant), Article B (blood sample of child), and Article C (blood sample of the suspect).

5 Procedure

1. DNA was extracted from the liquid blood samples as per the protocol described in Chap. 4 of the book.
2. Extracted DNA was quantified using the protocol described in Chap. 18.
3. Amplification of autosomal STRs was performed using the protocol described in Chap. 19.
4. Capillary electrophoresis/genotyping and subsequent data analysis were performed using the protocols described in Chaps. 23 and 24, respectively.

6 Results and Observation

1. Quantification of extracted DNA samples yielded the results as given in Table 26.1. The samples were further diluted using TE buffer to obtain the optimum DNA concentration for multiplex PCR.

Table 26.1
Estimation of DNA concentration of the extracted blood samples

	Article A	Article B	Article C
DNA content	200 ng/μl	150 ng/μl	350 ng/μl
Dilutions	1:200	1:150	1:350
Final concentration	1 ng/μl	1 ng/μl	1 ng/μl

2. The Punnett square table of the obtained autosomal STR DNA profile was as given in Table 26.2. Amelogenin marker shows "XY" suggesting that the questioned child is a male child. Further, the presence of maternal allele (highlighted) was found in the DNA profile of the questioned child. Similarly, the obligate paternal allele (underlined) was also found to be present in the DNA profile of the suspect. It suggests that the alleged father is the biological father of the questioned child. Statistical analysis can be performed to deduce the paternity index in this case.

7 Precautions

- Label all the tubes containing samples carefully.
- Handle the samples carefully, so that they do not get mixed up in the subsequent steps of experiment.
- Match the alleles one by one among the obtained profiles.
- Maintain the chain of custody for the samples.
- Maintain your internal laboratory record for case opening, DNA extraction, quantification, genotyping, and result interpretation.
- Refer to other precautions described in the individual experiments for DNA extraction (Chap. 4), quantification (Chap. 18), multiplex PCR (Chap. 19), genotyping (Chap. 23), and result interpretation (Chap. 24).

8 Troubleshooting

Refer to corresponding troubleshooting of the respective experiments used in this case study, i.e., DNA extraction (Chap. 4), quantification (Chap. 18), multiplex PCR (Chap. 19), genotyping (Chap. 23), and result interpretation (Chap. 24).

Table 26.2
Punnett square table of the samples analyzed

Genetic markers	Article A (complainant)	Article B (newborn)	Article C (suspect)
D3S1358	**15**,17	**15**,**15**	**15**,18
D1S1656	**11**,11	**11**,**14**	13,**14**
D2S441	**10**,10	**10**,**11**	**11**,14
D10S1248	**13**,16	**13**,**14**	**14**,14
D13S317	**9**,12	**9**,**12**	8,**12**
Penta E	7,**15**	**12**,**15**	**12**,19
D16S539	**8**,11	**8**,**11**	**11**,11
D18S51	14,**14**	**13**,**14**	**13**,14
D2S1338	22,**23**	**21**,**23**	17,**21**
CSF1PO	9,**10**	**10**,**12**	12,**12**
Penta D	**11**,12	**11**,**13**	**13**,14
TH01	**6**,9	**6**,**9**	7,**9**
vWA	**16**,16	**16**,**16**	**16**,17
D21S11	29,**30**	**30**,**33.2**	32.2,**33.2**
D7S820	**10**,12	**10**,**11**	**11**,11
D5S818	**11**,**12**	**11**,**12**	**11**,11
TPOX	**8**,10	**8**,**11**	**11**,11
D8S1179	10,**13**	**13**,**15**	14,**15**
D12S391	17,**20**	**19**,**20**	17,**19**
D19S433	**13**,13.2	**13**,**14.2**	**14.2**,16.2
SE33	30.2,**33.2**	**16**,**33.2**	**16**,16
D22S1045	15,**16**	**11**,**16**	**11**,11
FGA	**21**,24	**21**,**22**	21,**22**
Amelogenin	**X**,X	**X**,**Y**	X,**Y**
DYS391	–	**10**	**10**
DYS576	–	**19**	**19**
DYS570	–	**19**	**19**

Chapter 27

Solving a Case of Murder by DNA Fingerprinting: A Case Study

Objective: To solve a case of murder by DNA fingerprinting.

1 Introduction

Murder cases can be solved by DNA fingerprinting examination of the biological evidences collected during investigation. Based on the Locard's exchange principle, i.e., "with contact between two items, there will be an exchange of materials," evidences can be collected and referred for DNA examination. Evidences collected from the deceased, suspect, crime scene, murder weapon, as well as the witness present (if any) can be used for this purpose. Due to the requirement of low amount of DNA samples in the currently practiced DNA fingerprinting technology, substances as tiny as cigarette butt or strands of hairs are sufficient enough to establish a crime.

Certain evidences such as eye witnesses can lose credibility over time; however, the evidences generated by DNA examination have the capability to reveal truth even after 20 years of the commitment of offense. Some of the previously clueless evidences such as blood left on a broken window, saliva found on the cigarette butt or bear bottle, and skin cells on the wheels of a theft car are being solved nowadays with its full potential using DNA technology. Additional considerations need to be taken care of to minimize the degradation of evidences as well as to avoid contamination-related issues. Thus, the on-field analyst should be trained properly regarding the DNA evidences as well as their collection and preservation techniques. With the advancement of DNA technology and its requirement of a very less input of DNA quantity, many secondary evidences present in the crime scene can also be targeted such as baseball bat, hat or mask, eyeglasses, dirty laundry, toothpick, cigarette butt, envelope seal, bottle/glass, bite marks, as well as fingernail scrapings.

H. R. Dash et al., *Principles and Practices of DNA Analysis: A Laboratory Manual for Forensic DNA Typing*, Springer Protocols Handbooks, https://doi.org/10.1007/978-1-0716-0274-4_27, © Springer Science+Business Media, LLC, part of Springer Nature 2020

2 Principle

For solving a case of murder using DNA technology, the following considerations need to be followed:

(a) Deposition of suspect's DNA on the victim's body or clothing
(b) Deposition of suspect's DNA on any object
(c) Deposition of suspect's DNA on a location
(d) Deposition of victim's DNA on the suspect's body/clothing
(e) Deposition of victim's DNA on the object
(f) Deposition of victim's DNA on a location
(g) Deposition of witness's DNA (if any) on the victim/suspect
(h) Deposition of witness's DNA (if any) on an object or location

Accordingly biological evidences may be collected, preserved, and examined to link between the crime scene, victim/deceased, murder weapon, and witness (if any). In this regard, swab from the blood-stained floor or blood-stained soil can be collected from the crime scene along with the suitable controls. To link the deceased, clothing of the deceased or its body parts and bone samples for skeletonized body should be preserved. A murder weapon can be any hard, sharp, or blunt object capable of causing harm to the deceased. It may be recovered from the crime scene or from the memorandum of the suspect or from the body of the deceased. The common murder weapons may involve the stones/bricks, knife/sickle, or bullet. During autopsy bullets may be recovered from the body of the deceased followed by its DNA examination and control matching with the gun recovered from the suspect can link the suspect to the crime. Similarly, linking the suspect to the crime is of utmost importance in a case of murder. In this regard, DNA technology becomes highly useful by examining the clothing of the suspect with suspected bloodstains; rooted hairs found on the deceased body or in the crime scene; presence of cigarette butts, bear bottles, or glasses at the crime scene; or any other article found in the crime scene useful to link the suspect.

3 Brief Case History

A dermatologist's wife was stabbed to death by the suspect accusing of improper treatment by the doctor. Later when the assailant denied the crime, the investigation team took help of the DNA examination to corroborate the crime. In this case, the laboratory received clothing of the deceased, bloodstain swab from the floor, clothing of the suspect, and the knife for examination. The preliminary examination of the body fluid identification shows the presence of bloodstains in all the articles. Further they were subjected to DNA fingerprinting examination.

4 Samples

Clothing of the deceased (Article A), bloodstain swab from the floor (Article B), control swab (Article C), clothing from the suspect (Article D), knife seized from the suspect (Article E), and blood sample of suspect (Article F) were received in this case. All the samples were maintained in their recommended storage condition. Appropriate seal and label were checked properly prior to their examination.

5 Procedure

1. DNA was extracted from the deceased's clothing (Article A), swab of the floor (Article B), control swab (Article C), suspect's cloth (Article D), and knife (Article E) as per the protocol described in Chap. 5 of the book.
2. DNA was extracted from the blood sample of suspect (Article F) as per the protocol described in Chap. 4 of the book.
3. Extracted DNA was quantified using the protocol described in Chap. 18.
4. Amplification of autosomal STRs was performed using the protocol described in Chap. 19.
5. Capillary electrophoresis/genotyping and subsequent data analysis were performed using the protocols described in Chaps. 23 and 24, respectively.

6 Results and Observation

1. Quantification of extracted DNA samples yielded the results as given in Table 27.1. The samples were further diluted using TE buffer to obtain the optimum DNA concentration for multiplex PCR.
2. The results of the autosomal STR DNA profiling of the exhibits are given in Table 27.2. Amelogenin marker of Article A shows

Table 27.1
Estimation of DNA concentration of the extracted blood samples

	Article A	Article B	Article C	Article D	Article E	Article F
DNA content	100 ng/μl	75 ng/μl	0 ng/μl	25 ng/μl	12.5 ng/μl	175 ng/μl
Dilutions	1:100	1:75	–	1:25	1:12.5	1:175
Final concentration	1 ng/μl	1 ng/μl	–	1 ng/μl	1 ng/μl	1 ng/μl

Table 27.2
Autosomal STR DNA profile obtained from various examined articles

Genetic markers	Article A Clothing (deceased)	Article B Swab (floor)	Article C Swab (control)	Article D Clothing (suspect)	Article E Knife (suspect)	Article F Blood (suspect)
D3S1358	15,18	15,18	DNA profile could not be detected	14,**15**,16,**18**	15,18	14,16
D1S1656	13,16	13,16		**13,16**	13,16	16,16
D2S441	11,11	11,11		10,**11**	11,11	10,10
D10S1248	14,14	14,14		**14,14**	14,14	14,14
D13S317	8,8	8,8		**8,8**	8,8	8,8
Penta E	10,14	10,14		**10**,12,**14**,16	10,14	12,16
D16S539	9,12	9,12		8,**9,12**	9,12	8,12
D18S51	16,19	16,19		13,14,**16,19**	16,19	13,14
D2S1338	18,20	18,20		**18,20**,23,25	18,20	23,25
CSF1PO	12,12	12,12		10,11,**12**	12,12	10,11
Penta D	10,13	10,13		**10**,11,**13**	10,13	10,11
TH01	6,9	6,9		**6**,8,**9**	6,9	8,9
vWA	14,15	14,15		**14,15**	14,15	14,14
D21S11	30,34.2	30,34.2		28,**30**,31,**34.2**	30,34.2	28,31
D7S820	10,10	10,10		9,**10**,11	10,10	9,11
D5S818	11,11	11,11		**11**,12	11,11	11,12
TPOX	11,11	11,11		8,9,**11**	11,11	8,9
D8S1179	10,15	10,15		**10,15**	10,15	15,15
D12S391	18,20	18,20		**18,20**,24	18,20	20,24
D19S433	15.2,15.2	15.2,15.2		13,**15.2**	15.2,15.2	13,13
SE33	21,21.2	21,21.2		17,**21,21.2**,31.2	21,21.2	17,31.2
D22S1045	11,16	11,16		**11,16**	11,16	11,11
FGA	22,26	22,26		20,**22**,25.2,**26**	22,26	20,25.2
Amelogenin	X,X	X,X		**X**,Y	X,X	X,Y
DYS391	–	–		10	–	10
DYS576	–	–		18	–	18
DYS570	–	–		19	–	19

"XX" suggesting the deceased to be a female individual. No DNA profile was detected from the Article C suggesting no experimental error. Further, consistent autosomal STR DNA profile was detected from the source of Article A, Article B, and Article E. Article D yielded a mixed profile. All the alleles found from the victim's source are present in the mixed DNA profile obtained from the source of suspect's clothing, i.e., Article D (alleles highlighted and underlined). This suggests the transfer of body fluid of the deceased to the clothing of suspect during assault. Similarly, the alleles present in the DNA profile of the suspect's blood are also found to be present in the mixed DNA profile which mostly arises due to the regular wearing of the clothing. Further, statistical analysis can be performed for mixture interpretation as well as calculation of likelihood ratio between two DNA profiles.

7 Precautions

- Label all the tubes containing samples carefully.
- Handle the samples carefully, so that they do not get mixed up in the subsequent steps of experiment.
- Match the alleles one by one among the obtained profiles.
- Maintain the chain of custody for the samples.
- Maintain your internal laboratory record for case opening, DNA extraction, quantification, genotyping, and result interpretation.
- Refer to other precautions described in the individual experiments for DNA extraction (Chap. 4), quantification (Chap. 18), multiplex PCR (Chap. 19), genotyping (Chap. 23), and result interpretation (Chap. 24).

8 Troubleshooting

Refer to corresponding troubleshooting of the respective experiments used in this case study, i.e., DNA extraction (Chaps. 4 and 5), quantification (Chap. 18), multiplex PCR (Chap. 19), genotyping (Chap. 23), and result interpretation (Chap. 24).

Chapter 28

Solving a Case of Sexual Assault by DNA Fingerprinting: A Case Study

Objective: To solve a case of sexual assault by DNA fingerprinting.

1 Introduction

Sexual assault cases are more prevalent than any other types of crimes happening throughout the globe. Though sexual assaults are of many types, penetration of a body cavity is the most common form of assault. Intimacy of this assault leads to the transfer/exchange of body fluids without the presence of any common eyewitnesses. In many instances, the victim may be unable to provide the details of the perpetrator for varied reasons. In such instances, only DNA fingerprinting analysis becomes the masterpiece among scientific evidences to reach at the culprit.

In sexual assault cases, the timing of filing the complaint, medicolegal examination, and collection of evidences play important roles in establishing the crime by DNA examination. The medical examination of both the victim and assault should be carried out as per the local ethical guidance after obtaining the respective consents for the same. The medicolegal certificate should contain certain information regarding the details of the allegation as well as the medical and sexual history of the persons examined. With the increase in difference in time between the assault and collection of samples, both mechanical (due to drainage and hygiene) and biological elimination (degradation) of the evidences from the body takes place in addition to the physiologic dilution. Many samples from the victim may be collected based on the history of the assault such as anal, oral, or vaginal assault. Additionally, reference samples of both the suspect and the victim are required for DNA fingerprinting analysis.

H. R. Dash et al., *Principles and Practices of DNA Analysis: A Laboratory Manual for Forensic DNA Typing*, Springer Protocols Handbooks, https://doi.org/10.1007/978-1-0716-0274-4_28,

In earlier scenario identification microscopy, comparative microscopy, serological analyses, and biochemical analyses were performed for the biological samples. However, with the advent of DNA fingerprinting technology, it has replaced the conventional techniques due to its individualization nature. DNA fingerprinting examination of a sexual assault case reveals three types of results, i.e., either inclusion, exclusion, or no result. In all instances, the result becomes conclusive to determine the accused as well as to exonerate the falsely implicated suspect.

2 Principle

DNA of both the accused and victim is supposed to be present in the questioned samples collected in sexual assault cases. Due to the transfer of body fluids, the sexual assault cases can be solved in either of the two ways, i.e., detection of suspect's DNA profile on victim's belongings or detection of victim's DNA profile on belongings of the accused. The latter scenario occurs mostly in sexual assault cases of prepuberty victims. Sexual assault in most of the prepuberty victims results in profuse bleeding from the private parts resulting in the transfer of blood in the accused private parts in contact. Thus, DNA analysis from the clothing, pubic hair, penile, and/or scrotal swab may generate the victim's DNA profile confirming the sexual assault. However, in sexual assault cases of victims of postpuberty age, the common practice is to detect the DNA profile of the accused on the victim's belongings. In such cases, the common questioned articles include vagina/cervical swab, pubic hair, and clothing of the victim.

In all instances, a mixed autosomal STR DNA profile was expected in the questioned samples. As the current analysis involves the PCR process, preferential amplification of the DNA occurs. The articles from the victim's belongings are expected to contain more amount of victim's DNA in comparison to the DNA of the accused. Hence, after PCR process, a large number of DNA fragments of the victim's source mask the amplified DNA fragments of the accused. Subsequently only DNA from the victim's source is detected after genotyping. To overcome this problem, Y chromosome STR analysis is recommended which gives amplification of male DNA only present in the mixed sample. Most of the commercially available kits can successfully generate a male DNA profile when male-to-female DNA ratio is present as low as 1:1000. Additionally, due to statistical limitations and chance matching, sexual assault cases involving more than one accused cannot be interpreted by autosomal STR analysis. In such a scenario only Y chromosome STR analysis proves useful.

3 Brief Case History

A working lady in her 20s alleged that her manager sexually harassed her when working for a firm for over 1 year. She also alleged that her complaint to the management of the firm was continuously ignored. As per her allegation, she was denied promotion due to her complaint. After repeated complaint and continuous sexual assault, she resigned from her job and complained in the nearby police station. As per the victim, she was assaulted 1 day before her complaint at police station. As soon as the complaint was lodged, the victim was undergone medical test and the medical surgeon collected vaginal swab, undergarments, and pubic hair from her. The reference blood samples from both the victim and the suspect were also collected. Body fluid examination showed the presence of spermatozoa in the vaginal swab and undergarments of the victim, whereas the spermatozoa could not be detected from the pubic hair of the victim. Further all the articles were subjected to DNA fingerprinting examination.

4 Samples

Vaginal swab (Article A), control swab (Article B), pubic hair (Article C), undergarment (Article D), and blood sample (Article E) were collected from the victim. Similarly, reference blood sample (Article F) was collected from the suspect. All the samples were maintained in their recommended storage condition. Appropriate seal and label were checked properly prior to their examination.

5 Procedure

1. DNA was extracted from the victim's vaginal swab (Article A), control swab (Article B), pubic hair (Article C), and undergarment (Article D) as per the protocol described in Chap. 10 of the book.
2. DNA was extracted from the blood sample of victim (Article E) and suspect (Article F) as per the protocol described in Chap. 4 of the book.
3. Extracted DNA was quantified using the protocol described in Chap. 18.
4. Amplification of Y chromosome STRs was performed using the protocol described in Chap. 20.
5. Amplification of autosomal STRs was performed using the protocol described in Chap. 19.

6. Capillary electrophoresis/genotyping and subsequent data analysis were performed using the protocols described in Chaps. 23 and 24, respectively.

6 Results and Observation

1. Quantification of extracted DNA samples yielded the results as given in Table 28.1. The samples were further diluted using TE buffer to obtain the optimum DNA concentration for multiplex PCR.
2. The results of the Y chromosome STR DNA profile are given in Table 28.2. All the samples yielded a Y profile except Article B and Article E confirming no experimental error. No male DNA profile was detected from the source of Article C. Further, same/consistent Y chromosome STR DNA profile was detected from the source of Article A, Article D, and Article F.
3. The results of autosomal chromosome STR DNA profile are given in Table 28.3. No autosomal STR DNA profile could be detected from the source of Article B confirming no experimental error. Same/consistent female autosomal STR DNA profile was obtained from the source of Articles C and E suggesting their same origin. Additionally, mixed autosomal STR DNA profile was obtained from the source of Article A and Article D. All the alleles generated from the suspect's blood sample (Article F) were found to be included in the mixed DNA profile obtained from Article A and Article D. Similarly, all the alleles generated from the victim's blood sample (Article E) were also found to be included in the mixed DNA profile obtained from Article A and Article D. Further statistical analysis can be performed for mixture interpretation.

7 Precautions

- Label all the tubes containing samples carefully.
- Handle the samples carefully, so that they do not get mixed up in the subsequent steps of experiment.

Table 28.1
Estimation of DNA concentration of the extracted blood samples

	Article A	Article B	Article C	Article D	Article E	Article F
DNA content	350 ng/μl	–	15 ng/μl	200 ng/μl	450 ng/μl	500 ng/μl
Dilutions	1:350	–	1:15	1:200	1:450	1:500
Final concentration	1 ng/μl	–	1 ng/μl	1 ng/μl	1 ng/μl	1 ng/μl

Table 28.2
Y chromosome STR DNA profile obtained from various examined articles

Genetic markers	Article A Vaginal swab (victim)	Article B (Control swab)	Article C Pubic hair (victim)	Article D Undergarment (victim)	Article E Blood (victim)	Article F Blood (suspect)
DYS576	19	Y-STR DNA profile could not be detected	Y-STR DNA profile could not be detected	19	Y-STR DNA profile could not be detected	19
DYS389I	14			14		14
DYS635	21			21		21
DYS389II	32			32		32
DYS627	21			21		21
DYS460	10			10		10
DYS458	16			16		16
DYS19	15			15		15
YGATAH4	11			11		11
DYS448	18			18		18
DYS391	10			10		10
DYS456	16			16		16
DYS390	25			25		25
DYS438	10			10		10
DYS392	13			13		13
DYS518	43			43		43
DYS570	15			15		15
DYS437	14			14		14
DYS385	15,20			15,20		15,20
DYS449	33			33		33
DYS393	14			14		14
DYS439	11			11		11
DYS481	23			23		23
DYF387S1	36,37			36,37		36,37
DYS533	10			10		10

- Match the alleles one by one among the obtained profiles.
- Maintain the chain of custody for the samples.
- Maintain your internal laboratory record for case opening, DNA extraction, quantification, genotyping, and result interpretation.

Table 28.3
Autosomal STR DNA profile of the examined articles

Genetic markers	Article A Vaginal swab (victim)	Article B (Control swab)	Article C Pubic hair (victim)	Article D Undergarment (victim)	Article E Blood (victim)	Article F Blood (suspect)
D3S1358	15,16,17	DNA profile could not be detected	15,16	15,16,17	15,16	16,17
D1S1656	8,12,13,15		12,13	8,12,13,15	12,13	8,15
D2S441	10,11,13		10,11	10,11,13	10,11	10,13
D10S1248	13,14,15,16		13,15	13,14,15,16	13,15	14,16
D13S317	8,10,13		10,13	8,10,13	10,13	8,10
Penta E	7,12,13,19		7,13	7,12,13,19	7,13	12,19
D16S539	10,11,13		10,11	10,11,13	10,11	11,13
D18S51	12,13,17		13,17	12,13,17	13,17	12,13
D2S1338	18,19,23,24		19,23	18,19,23,24	19,23	18,24
CSF1PO	11,12		11,11	11,12	11,11	12,12
Penta D	9,10,11,12		11,12	9,10,11,12	11,12	9,10
TH01	6,8,9		6,8	6,8,9	6,8	8,9
vWA	14,18		14,18	14,18	14,18	14,14
D21S11	28,29,30,32.2		28,29	28,29,30,32.2	28,29	30,32.2
D7S820	8,9,10,11		8,9	8,9,10,11	8,9	10,11
D5S818	11,13		11,11	11,13	11,11	11,13
TPOX	8,11		8,11	8,11	8,11	8,11
D8S1179	13,15,16		15,16	13,15,16	15,16	13,15
D12S391	18,20		18,20	18,20	18,20	18,18
D19S433	13,14		14,14	13,14	14,14	13,13
SE33	20,21.1,25.2,31.2		21.1,25.2	20,21.1,25.2,31.2	21.1,25.2	20,31.2
D22S1045	11,15		15,15	11,15	15,15	11,15
FGA	21,23,24		24,24	21,23,24	24,24	21,23
Amelogenin	X,Y		X,X	X,Y	X,X	X,Y
DYS391	10		–	10	–	10
DYS576	19		–	19	–	19
DYS570	18		–	18	–	18

- Refer to other precautions described in the individual experiments for DNA extraction (Chap. 4), quantification (Chap. 18), multiplex PCR (Chap. 19), genotyping (Chap. 23), and result interpretation (Chap. 24).

8 Troubleshooting

Refer to corresponding troubleshooting of the respective experiments used in this case study, i.e., DNA extraction (Chaps. 4 and 10), quantification (Chap. 18), multiplex PCR (Chap. 19), genotyping (Chap. 23), and result interpretation (Chap. 24).

Chapter 29

Identification of an Unknown Skeleton by DNA Fingerprinting: A Case Study

Objective: To identify an unknown skeleton by DNA fingerprinting.

1 Introduction

Identification of human remains has been transformed manyfold since last two decades due to the advent of this fingerprinting technology. Before the discovery of with technology, other techniques such as dactyloscopy, anthropology, odontology, and medicolegal examinations were used for identification purposes. These techniques provide limited information in comparison to the molecular level of DNA examination which determines the exact origin of the unknown human remains. The human remains may be of soft or degraded tissues, bones, teeth, or hairs from which the genetic information can be extracted. As the fundamental of genetics states that two close relatives share the genetic variants, the DNA profile of the unknown human remains can be matched with the DNA profile of the tentative relatives. Based on the result, conclusion on the identification of the unknown deceased can be drawn.

The identification of a missing person can also be carried out using the medical or personal objects. Medical samples including the biopsies, umbilical cords, teeth, or other parts of the body can provide the ideal samples for the identification of a missing person. Similarly, personal artifacts also provide simple and powerful matching of DNA. Personal belongings such as a hairbrush, toothbrush, and razor can be analyzed to match with the DNA profile of the missing person. Additionally, database-led identification programs can be conducted by matching the unknown profile with the database of population DNA profiles.

H. R. Dash et al., *Principles and Practices of DNA Analysis: A Laboratory Manual for Forensic DNA Typing*, Springer Protocols Handbooks, https://doi.org/10.1007/978-1-0716-0274-4_29,

2 Principle

Analysis of DNA in the case of identification of human remains is comprised of five different steps such as

(a) *Extraction of DNA from the human remains*

(b) *Extraction of DNA from the reference samples*

(c) *Generating the DNA profile from both the samples, i.e., human remains and reference sample*

(d) *Comparison of the DNA profiles of the questioned and the reference samples*

(e) *Conclusion on the basis of degree of matching between the questioned and reference samples*

Identification may be required in cases such as burning leading to death, army conflicts, recovery of skeletonized body, and recovery of body parts. In all the cases, the DNA starts to degrade post-death of the individual depending on the environment. For example, environments with warm and humid conditions increase the degradation rate, whereas the cold and dry environments help in preserving the DNA samples. However, in most of the cases, it becomes a huge challenge for the examiner to generate a DNA profile from the degraded samples. Thus, when human remains are recovered in a short time after death, soft-tissue samples such as muscle tissues can be preserved for DNA examination. When doubt rises regarding degradation of the soft tissues, hard tissues such as bones and teeth may be collected for identification purpose. As the cells present in the hard tissues are embedded in a dense biomineral matrix, they are least affected by decomposition as well as putrefaction. Though all bones preserve DNA, long bones such as femur are better sources of DNA so as the teeth.

Correct identification of a skeletal remain depends on the availability of reference samples. In this regard, presence of a relationship triangle, i.e., father-mother-son or father-mother-daughter, is the most suitable sample for conclusive identification of an individual by autosomal STR typing (Fig. 29.1). In reference

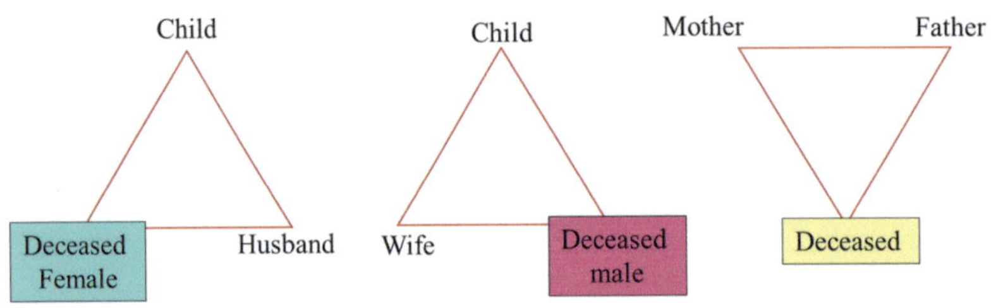

Fig. 29.1 Significance of a triangle of relationships for proper identification of the deceased

sample-limited conditions, patrilineal lineage of a missing person can be established by analyzing the Y chromosome profile. Similarly, mtDNA analysis can also determine the matrilineal lineage of an individual. Further, matching among the siblings can also be carried out; however, it may not identify an individual conclusively.

3 Brief Case History

A group of dacoits kidnapped an individual from his home in a remote village of India. After 3 days some local people observed some burnt skeletal remains inside the nearby jungle. They informed police and the autopsy surgeon confirmed the burnt skeletal remains to be of human origin. Investigating team suspected this to be the action of the group of dacoits. Due to the availability of only burnt skeletal remains, phenotypic identification of the deceased could not be performed. Hence, the half-burnt skeletal samples along with the reference blood samples of its parents were received for DNA examination.

4 Samples

Half-burnt bone samples (Article A) were collected after medical report confirming their human origin. Reference blood samples from the putative father (Article B) and the putative mother (Article C) were collected. All the samples were maintained in their recommended storage condition. Appropriate seal and label were checked properly prior to their examination.

5 Procedure

1. DNA was extracted from the blood samples (Article B and Article C) as per the protocol described in Chap. 4 of the book.
2. DNA was extracted from the bone samples (Article A) as per the protocol described in Chap. 9 of the book.
3. Extracted DNA was quantified using the protocol described in Chap. 18.
4. Amplification of autosomal STRs was performed using the protocol described in Chap. 19.
5. Capillary electrophoresis/genotyping and subsequent data analysis were performed using the protocols described in Chaps. 23 and 24, respectively.

Table 29.1
Estimation of DNA concentration of the extracted blood samples

	Article A	Article B	Article C
DNA content	3 ng/μl	350 ng/μl	200 ng/μl
Dilutions	1:3	1:350	1:200
Final concentration	1 ng/μl	1 ng/μl	1 ng/μl

6 Results and Observation

1. Quantification of extracted DNA samples yielded the results as given in Table 29.1. The samples were further diluted using TE buffer to obtain the optimum DNA concentration for multiplex PCR.
2. The results of autosomal chromosome STR DNA profile are given in Table 29.2.
3. Amelogenin marker shows "XY" suggesting the deceased to be a male individual. Further, the maternal allele (highlighted) was found to be present in the DNA profile of the skeletal remains. Similarly, the obligate paternal allele (underlined) was also found to be present in the DNA profile of the putative father. It suggests that the skeletal remains belong to the biological offspring of the putative father and the mother. Further, statistical analysis can be performed to deduce the paternity index in this case.

7 Precautions

- Label all the tubes containing samples carefully.
- Handle the samples carefully, so that they do not get mixed up in the subsequent steps of experiment.
- Match the alleles one by one among the obtained profiles.
- Maintain the chain of custody for the samples.
- Maintain your internal laboratory record for case opening, DNA extraction, quantification, genotyping, and result interpretation.
- Refer to other precautions described in the individual experiments for DNA extraction (Chap. 4), quantification (Chap. 18), multiplex PCR (Chap. 19), genotyping (Chap. 23), and result interpretation (Chap. 24).

Table 29.2
Autosomal STR DNA profile of the examined articles

Genetic markers	Article C Blood sample (putative mother)	Article A Skeletal remains (deceased)	Article B Blood sample (putative father)
D3S1358	15,**16**	**14,16**	**14**,15
D1S1656	**15**,16.3	**15,15**	**15**,16
D2S441	10,**14**	**10,14**	**10**,10
D10S1248	16,**16**	**16,16**	**16**,16
D13S317	11,**11**	**11,11**	**11**,12
Penta E	15,**21**	**11,21**	**11**,11
D16S539	**12**,13	**10,12**	9,**10**
D18S51	**10**,14	**10,15**	11,**15**
D2S1338	**18**,25	**18,21**	**21**,25
CSF1PO	11,**12**	**12,12**	11,**12**
Penta D	**9**,13	**9,9**	**9**,11
TH01	**6**,10	**6,9.3**	**9.3**,9.3
vWA	**18**,20	**17,18**	14,**17**
D21S11	29,**30**	**30,31.2**	30,**31.2**
D7S820	8,**8**	**8,8**	**8**,12
D5S818	**11**,12	**11,12**	11,**12**
TPOX	**11**,11	**8,11**	**8**,11
D8S1179	**14**,16	**14,15**	11,**15**
D12S391	20,**21**	**18,21**	**18**,18
D19S433	**12.2**,16.2	**12.2,13**	**13**,15
SE33	19.3,**32.2**	**30.2,32.2**	19,**30.2**
D22S1045	15,**15**	**15,15**	**15**,15
FGA	**21**,24	**21,24**	20,**24**
Amelogenin	X,**X**	**X,Y**	X,**Y**
DYS391	–	**10**	**10**
DYS576	–	**15**	**15**
DYS570	–	**16**	**16**

8 Troubleshooting

Refer to corresponding troubleshooting of the respective experiments used in this case study, i.e., DNA extraction (Chaps. 4 and 5), quantification (Chap. 18), multiplex PCR (Chap. 19), genotyping (Chap. 23), and result interpretation (Chap. 24).

Part VIII

Advanced Technologies for Forensic DNA Analysis

Chapter 30

Sequencing Control Region of Human Mitochondrial Genome to Assess Matrilineal Lineage

Objective: To analyze control region of mitochondria by using Sanger sequencing technique.

1 Introduction

Mitochondrion is the exclusive cell organelle of human cells which contains extracellular DNA. Uniqueness of mitochondrial DNA (mtDNA) involves its double-stranded circular and histone-free structure. The compact mtDNA codes for 13 polypeptides involved in oxidative phosphorylation as well as 2 rRNAs and 22 tRNAs. Besides, it contains a control region of around 1100 bp length which is noncoding in nature. The copy number of mtDNA varies from 2 to 10 per mitochondrion whereas a cell contains as many as 1000 mitochondria depending on the physiological condition of the cell (Budowle et al. 2003). However, size of mtDNA is quite smaller than the genomic DNA as the approximate size of genomic DNA is around 3 billion bp, whereas the size of circular mitochondrial DNA is 16,569 bp (Lutz et al. 2000).

The unique inheritance pattern of mtDNA advocates its widespread application in anthropology, evolution, and forensic analyses. Unlike to the nuclear DNA, mtDNA is inherited exclusively from the mother with little or no contribution from the father. Due to uniparental inheritance, mtDNA does not undergo recombination; hence it represents matrilineal lineage in generations. From forensic point of view, mtDNA analysis is of much importance as they are found in multiple copies per cell contributing to higher amount of DNA in comparison to nuclear DNA, mostly in compromised forensic samples. Robustness and resistance to extreme environmental conditions make mtDNA useful for ancient DNA

H. R. Dash et al., *Principles and Practices of DNA Analysis: A Laboratory Manual for Forensic DNA Typing*, Springer Protocols Handbooks, https://doi.org/10.1007/978-1-0716-0274-4_30,

analysis. Additionally, mtDNA analysis is useful in samples such as hair shaft, where chance of getting nuclear DNA result is negligible.

mtDNA analysis has a wide variety of applications in history. Victims of world trade center were identified using mtDNA sequencing technology. Analysis of remains of a Neanderthal man (Krings et al. 1997), World War victims (Pajnič et al. 2010), and a 7000-year-old brain tissue (Paabo et al. 1988) through mtDNA sequencing has reflected the utility of this technique. However, mtDNA sequencing technology earned a huge appreciation throughout the globe in case of verification of Tsar Nicholas II's bones. mtDNA sequence was successfully compared with the living maternal relatives and putative bone remains of Tsar Nicholas II were identified (Gill et al. 1994). However, the analysis procedure of mtDNA varies from the routine nuclear DNA analysis which has been described in details in this chapter.

2 Principle

The major goal of forensic DNA analysis is to differentiate among individuals. In this context, the control regions of mtDNA which are highly polymorphic in nature become the obvious choice for analysis. In general, two hypervariable regions are present in mtDNA called as HVRI and HVRII which show genomic variation at the highest level. As these regions do not contain any coding gene sequences, their mutation rate becomes ten times higher than that of the corresponding coding regions. The length of HVRI region (342 bp) is more than the HVRII region (268 bp) (Fig. 30.1). In some instances, a small (137 bp) third hypervariable region (HVRIII) is also present in human mtDNA which is useful in resolving samples where HVRI/HVRII analysis does not provide sufficient differentiation.

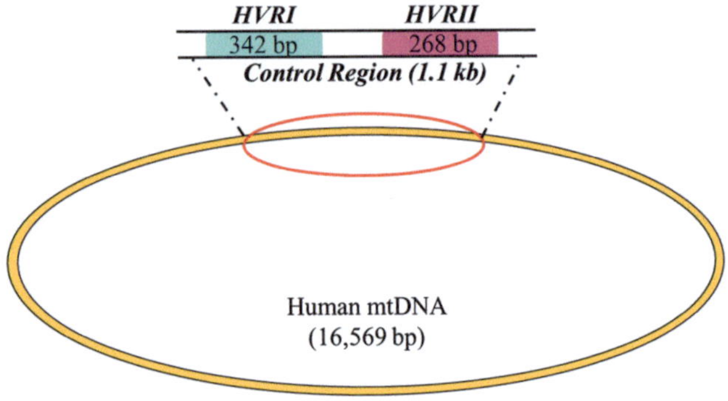

Fig. 30.1 An overview of human mitochondrial DNA with specific reference to control region

For analysis, the mtDNA is extracted from the biological samples and the target region; most preferably the hypervariable regions are amplified using the primer sets followed by sequencing using BigDye Terminator Cycle Sequencing kit (Thermo Fisher Scientific). During this process the conventional chain termination technique developed by Frederick Sanger is used. This technique involves amplification of the target region by routine PCR method using dNTPs as well as labeled ddNTPs. When ddNTPs are incorporated into the strand during amplification process, termination of template amplification occurs and different DNA fragments are generated. Thus, finally, DNA fragments of different sizes each ending with a different nucleotide labeled with different dye are generated. Finally, capillary electrophoresis is performed and the software analyzes the sequence of a DNA strand based on the size of the fragment and the emitted fluorescence.

The resultant DNA sequences are analyzed using specific analysis software and the sequence similarities and dissimilarities are calculated. After obtaining mtDNA sequence, it is first compared with the Cambridge or Anderson reference sequence or the revised Cambridge sequence (rCRS). Based on the sequence similarities or dissimilarities, the result can be interpreted as either inclusion, exclusion, or inconclusive. As per Scientific Working Group on DNA Analysis (SWGDAM)'s recommendation, difference of two or more nucleotide sequences between the questioned and reference samples can be referred as exclusion. However, when both the sequences are exactly same, they cannot be excluded of having the same origin. An inconclusive result is obtained when there exists a single-nucleotide difference between the questioned and reference sample. Few mtDNA population databases randomly selecting different individuals are present which include FBI mtDNA database, EMPOP (http://www.empop.org), and mtDNA manager (http://mtmanager.yonsei.ac.kr/).

3 Reagents/Materials Required and Their Role

3.1 Template DNA

Specific DNA extraction protocol should be followed for extraction of mtDNA from degraded forensic samples. Routine DNA quantification steps may not solve the purpose as they estimate the total DNA including microbial DNA and in some instances total human DNA including nuclear and mitochondrial DNA.

3.2 Primers

For amplification of whole control region of mitochondrial DNA, use the following primer sets, i.e., L15977: 5′ CAC CAT TAG CAC CCA AAG CT 3′ (Ward et al. 1991) and H599: 5′ TTG AGG AGG TAA GCT ACA TA 3′ (Brandstätter et al. 2004). If a high quality of mtDNA is extracted from the forensic samples, the entire control region is amplified using this primer set which is approximately

1.1 kb in size. If a poor quality of mtDNA is extracted an alternative strategy known as Midi-Mito strategy is applied amplifying the control region in multiple overlapping fragments. During sequencing seven sequencing reactions are performed using primers L15977, H16175, H16401, L16450, H274, L314, and H599.

3.3 MyTaq™ HS Red Mix 2×

MyTaqTM HS Red Mix (BIOLINE) is the ready-to-use mixture for setting up of hot-start PCR which contains the Taq DNA polymerase and dNTPS required for the PCR process. The mixture is powered by the antibody-mediated hot-start technique which does not allow the nonspecific amplification during the reaction setup. It is recommended to be stored at −20 °C and repeated freeze-thaw should be avoided to ensure optimum activity of the reagents.

3.4 illustra™ ExoProStar™

It is the enzymatic technique used for PCR and sequencing reaction cleanup. It is a double-enzyme reaction mechanism to remove leftover primers and dNTPs used during a PCR process. The reaction simultaneously uses alkaline phosphatase and exonuclease I. Exonuclease I digests single-stranded DNA to release deoxyribonucleoside 5′ monophosphates (dNMPs) whereas alkaline phosphatase catalyzes the dephosphorylation of unincorporated nucleotides and dNMPs to release nucleosides and Pi.

3.5 ABI Prism® BigDye™ Terminator v3.1 Cycle Sequencing Kit

The kit provides useful chemicals and reagents required for sequencing a DNA fragment. The kit is useful for sequencing single- or double-stranded DNA templates. The components of the kit include ready reaction mix, dilution buffer, control DNA, and control primers. The ready reaction mix, control DNA, and primers should be stored at −15 to −25 °C whereas the sequence buffer can be stored at 4 °C.

3.6 Molecular Biology-Grade Water

Molecular biology-grade water is the form of ultrapure high-quality water which is free from DNase, RNase, and protease contamination. Additionally, no toxic agents such as DEPC should be present in this water. Milli-Q water can be produced in the laboratory or it can be procured from any molecular biology-grade chemical manufacturer.

3.7 Genetic Analyzer 3500xL

Genetic Analyzer 3500xL instrument is one of the advanced versions of sequencer which can accommodate 24 samples in a single run. The instrument relies on the principle of capillary electrophoresis where the conventional gel slab has been replaced with thin capillaries and electrophoresis takes place. The only difference is that the result of capillary electrophoresis is being generated in the form of peaks in comparison to the bands of the slab gel electrophoresis result. The dedicated data collection software analyzes the peaks and designates nucleotides to the peaks to generate a complete DNA sequence.

Table 30.1
Recommended amounts of reagents required to prepare amplification mixture for PCR

PCR components	Volume per reaction (μl)	No. of reactions	Total volume
MyTaq HS Red Mix 2×	10.0	A	10A
L15977 primer (5 μM)	0.8	A	0.8A
H599 primer (5 μM)	0.8	A	0.8A
Molecular biology-grade water	6.4	A	6.4A
Total volume			*(10 + 0.8 + 0.8 + 6.4)A*

4 Procedure

1. Calculate the requirement of each component of the PCR process using the following table as per your requirement/ number of samples (Table 30.1).
2. Thaw all the PCR components in ice, vortex for 3–5 s, and centrifuge briefly before using.
3. Pipette out the calculated amount of water, MyTaq HS Red Mix, and primer sets in a 1.5 ml microcentrifuge tube marked as Amplification Mixture (AM). It is always recommended to add water first to the tube, followed by MyTaq HS Red Mix and primer sets.
4. As soon as the preparation of AM is over, immediately store the tubes containing PCR components in the recommended storage condition.
5. Vortex the amplification mixture for 5–10 s, and then dispense (10 + 0.8 + 0.8 + 6.4) μl of amplification mixture to each 200 μl PCR tube pre-marked for each sample.
6. Add 2.0 μl of the extracted mtDNA to each designated PCR tube containing the amplification mixture.
7. In one tube add control DNA (2800 M DNA) in the amplification mix and in another tube add molecular biology-grade water or TE buffer in place of template DNA which will be considered as positive and negative control, respectively.
8. Close the tubes properly and briefly centrifuge the tubes so that all its contents will be present at the bottom of the tube.
9. Place the tubes in the thermal cycler. Set the thermal cycling protocol as 95 °C for 3 min followed by 36 cycles of 94 °C for 15 s, 56 °C for 1 min, and 72 °C for 2 min, and finally hold at 4 °C for ∞.
10. The amplified products can be visualized at 1.5% agarose gel and under UV light.

11. When a proper amplification is evident from the agarose gel electrophoresis process, perform PCR cleanup using illustra™ ExoProStar™ PCR cleanup kit. Add 1.5 μl of ExoProStar™ 1 to the remaining 10–12 μl of PCR product and incubate the tube at 37 °C for 45 min followed by 80 °C for 15 min.
12. Perform separate sequencing reaction using ABI Prism® Big-Dye™ terminator cycle sequencing kit for each primer using the PCR product as template.
13. Calculate the amount of each reagent using Table 30.2 as per your requirement.
14. Vortex the amplification mixture for 5–10 s, and then dispense (0.68 + 1.02 + 0.408 + 3.6) μl of amplification mixture to the wells of a 96-well plate.
15. Add 1 μl of purified PCR product to the corresponding wells of the 96-well plate.
16. Seal the plate with a PCR seal.
17. Perform the sequencing reaction in a PCR machine using the protocol as 96 °C for 4 min followed by 25 cycles of 96 °C for 15 s, 50 °C for 10 s, and 60 °C for 2 min.
18. Add 1.7 μl of 125 mM EDTA and 20 μl of 100% ethanol to each well, vortex briefly, and incubate at room temperature for 15 min.
19. Centrifuge the plate at 2250 × *g* for 30 min to precipitate the sequencing products.
20. Discard the supernatants by inverting the plate and centrifuge at inverted position at 4000 rpm for 10 min.
21. Wash the plate twice using 70% ethanol by repeating steps 18 and 19.
22. Incubate the plate at room temperature till the leftover ethanol is evaporated.

Table 30.2
Recommended amounts of reagents required to prepare amplification mixture for sequencing PCR

PCR components	Volume per reaction (μl)	No. of reactions	Total volume
Ready reaction	0.68	A	0.68A
Dilution buffer	1.02	A	1.02A
Sequencing primer (5 μM) (L15977/ H16175/H16401/L16450/ H274/ L314/H599)	0.408	A	0.408A
Molecular biology-grade water	3.6	A	3.6A
Total volume			*(0.68 + 1.02 + 0.408 + 3.6)A*

23. Add 10 μl of deionized formamide to each well followed by heating at 95 °C for 3 min and snap cool in ice for 5 min.
24. Run the plate on the capillary electrophoresis instrument Genetic Analyzer 3500xL (Thermo Scientific).
25. Align the obtained sequences using the software SeqScape (Life Technologies) or Sequencer (Gene Codes).
26. Analyze the control region sequence by comparing with the revised Cambridge Reference Sequence (rCRS). Additionally, few points need to be considered while analyzing the obtained mtDNA sequences:
    (a) *Try to have a complete sequence coverage for the control region.*
    (b) *Observe the common phenomena of difference in sequence between the examined and reference mtDNA at position 263 and insertion of "C" following 315 position.*
    (c) *Consider length heteroplasmy while analyzing mtDNA sequence which generally arises due to insertion or deletion of a base.*
27. Use EMPOP (www.empop.org) to calculate the frequency of generated mtDNA sequence.

5 Observation and Result

Concordance: When two mtDNA sequences, i.e., one from the questioned sample and another from the reference sample, are consistent with each other, two samples cannot be excluded to be originated from the same matrilineal lineage.

Exclusion: When two mtDNA sequences, i.e., one from the questioned sample and another from the reference sample, differ by two or more nucleotides, the samples are excluded to be originated from the same matrilineal lineage.

Inconclusive: When two mtDNA sequences, i.e., one from the questioned sample and another from the reference sample, differ by only one nucleotide, the resulting comparison can be concluded to be inconclusive.

6 Revised Cambridge Reference Sequence for *HVRI*

16024: TTCTTTCATG GGGAAGCAGA TTTGGGTACC ACCCAAGTAT

16064: TGACTCACCC ATCAACAACC GCTATGTATT TCGTACATTA

16104: CTGCCAGCCA CCATGAATAT TGTACGGTAC CATAAATACT

16144: TGACCACCTG TAGTACATAA AAACCCAATC CACATCAAAA

16184: CCCCCTCCCC ATGCTTACAA GCAAGTACAG CAATCAACCC

16224: TCAACTATCA CACATCAACT GCAACTCCAA AGCCACCCCT

16264: CACCCACTAG GATACCAACA AACCTACCCA CCCTTAACAG

16304: TACATAGTAC ATAAAGCCAT TTACCGTACA TAGCACATTA

16344: CAGTCAAATC CCTTCTCGTC CC-16365

7 Revised Cambridge Reference Sequence for *HVRII*

73: ATGCACGCGA TAGCATTGCG AGACGCTGGA GCCGGAGCAC

113: CCTATGTCGC AGTATCTGTC TTTGATTCCT GCCTCATCCT

153: ATTATTTATC GCACCTACGT TCAATATTAC AGGCGAACAT

193: ACTTACTAAA GTGTGTTAAT TAATTAATGC TTGTAGGACA

233: TAATAATAAC AATTGAATGT CTGCACAGCC ACTTTCCACA

273: CAGACATCAT AACAAAAAAT TTCCACCAAA CCCCCCCTCC

313: CCCGCTTCTG GCCACAGCAC TTAAACAC-340(end)

8 Precautions

- Wear gloves and goggles during isolation of mitochondrial DNA.
- Always use disposable plasticware which are pre-sterilized and DNase free.
- Use filter tips to avoid cross-sample contamination.
- Do not store the reagents under a light source.
- Always include both positive and negative controls throughout the process.
- As some of the chemicals are hazardous, they should be handled carefully.
- Always prepare amplification mixture for 1–2 extra reactions to avoid any chance of possible pipetting error.

- Always store the reagents in appropriate storage conditions.
- Never use expired reagents.
- In between runs do not close the data collection software and open it. It may hamper the life span of the laser present in the instrument.
- Never mix the working area for nuclear DNA and mitochondrial DNA.

9 Troubleshooting

Problem	Tentative cause	Possible solutions
Mixed sequences (no mixed sequence in positive control and no sequences in negative control)	RNA contamination	Run the samples on agarose gel; check the quality of sample; if RNA is present at higher amount use RNase treatment or re-isolate the samples
		Check A260/A280 of the extracted DNA; if the observed value is greater than 2.0 go for RNase treatment or re-isolate the samples
	DNA contamination	Run on agarose gel to check the presence of two different bands; take extra precaution and re-isolate the sample
	Primer not optimized	Use recommended primer concentration; check the specificity of the primer before proceeding for sequencing
	Carryover contamination from septa	Replace septa with a fresh one and rerun the samples
	Leftover primers present in the product	Remove leftover PCR reagents using post-PCR purification kit

(continued)

Problem	**Tentative cause**	**Possible solutions**
Little or "no" sequence signals	Template quantity not adequate	Always recommended amount of DNA template
	Primer quantity not adequate	Use recommended concentration of primer during sequencing PCR
	Broken or blocked capillaries	Check and replace the defunct capillaries
	PCR process not complete	Check expiry of the PCR reagents and re-PCR the samples
Issue with data analysis	Threshold settings for quality and mixed base settings are not defined properly	Correct the settings and reanalyze the result
	Heterozygous insertion/deletion	Recheck the sequence on the reverse strand

Chapter 31

Application of Next-Generation Sequencing (NGS) in Forensic DNA Analysis and DNA Phenotyping

Objective: To analyze forensic samples by using next-generation sequencing (NGS).

1 Introduction

Use of DNA fingerprinting to exploit the variation of fragment length of STR markers has become an integral part of the conventional forensic DNA analysis. Mostly capillary electrophoresis technique is applied in such cases to determine the fragment length variation of STR markers present on X, Y, and autosomal chromosomes. In certain instances, certain regions on mitochondrial DNA (mtDNA) are also analyzed by conventional Sanger sequencing technology mostly to analyze the matrilineal inheritance. However, none of the currently used technique allows any additional lead to the investigation purpose without a reference sample or a hit in the database. In this context, after the completion of the Human Genome Project in 2003, the next-generation sequencing technique has become widely useful in conventional as well as forensic applications with the advanced features of massively parallel sequencing (MPS), high throughput, fastness, depth coverage, low cost per nucleotide, and reduction in error. With the increase in load of casework samples and requirement of establishment of criminal intelligence database throughout the globe, the NGS technique is widely useful to meet the current demand of forensic DNA typing.

In the conventional capillary electrophoresis technique, the difference in sizes of the individual markers is being assessed, whereas in targeted MPS both length and sequence variation can be measured for higher number of markers in a single run. Use of

H. R. Dash et al., *Principles and Practices of DNA Analysis: A Laboratory Manual for Forensic DNA Typing*, Springer Protocols Handbooks, https://doi.org/10.1007/978-1-0716-0274-4_31,

different dye sets limits the number of STR markers analyzed in a single multiplex reaction; however, each molecule is read independently enabling more number of markers to be analyzed simultaneously in a single MPS reaction. Thus, in contrary to the current capacity of CE technique to analyze 15–27 STR markers, a single MPS reaction can analyze ~400 markers of autosomal, Y, and X STRs as well as various single-nucleotide polymorphism (SNP) markers. Additionally, mtDNA can also be analyzed in this technique. Another advantage of the NGS technique over the conventional CE technique is the high sensitivity of the process.

Many NGS platforms are currently available from Roche, Ion torrent (Thermo Fisher Scientific), Illumina, Solid, PacBio, Helicose, Oxford Nanopore, Qiagen, and BGI (Sobiah et al. 2018). However, for forensic application two most common platforms being used include the MiSeq® FGx™ Forensic Genomics System (Illumina) and the HID-Ion Personal Genome Machine (PGM)™ (Thermo Fisher Scientific). Though the workflow varies a little between these two systems, both of them are capable of providing an enhanced solution for the smallest, most compromised, as well as mixed samples. The MiSeq® FGx™ Forensic Genomics System (Illumina) has been described in details in this chapter.

2 Principle

Amplicon-based sequencing is the basic principle behind an NGS workflow to be used for forensic application. The first step involves the library preparation where the library is prepared using a two-step amplification process. In the first amplification process the target sequence is amplified using sequence-specific primer pairs and the second amplification step involves addition of indexes and adaptors to give a unique identity to each amplicon. After preparation of library, they are pooled in one tube and linearized. Further, the library is injected into the flow cell where oligos complementary to the adaptors are present and the fragments are amplified by bridge amplification to generate clonal clusters. Finally, the sequencing by synthesis (SBS) technique is employed for sequencing which involves the detection of single bases during their incorporation into the DNA strands. Finally, the obtained sequence data are analyzed using the ForenSeq™ Universal Analysis Software to generate the result.

In the primary step, library is prepared using ForenSeq™ DNA Signature Prep Kit for simultaneous amplification of 231 markers (27 autosomal, 24 Y, 7 X, 95 identity SNPs, 22 phenotypic SNPs, and 56 ancestry SNPs) useful for forensic casework and database purpose. Besides autosomal, Y-, and X-STR markers, the kit contains identity informative single-nucleotide polymorphism (SNP) markers that are useful for source designation in case of degraded samples or samples with potential PCR inhibitors. Additionally, some phenotype SNPs informative for tentative eye and hair color of the individual as well as ancestry SNPs informative for biogeographical origin of the individual are also included in the kit. The analysis software automatically analyzes the obtained result leading to generation of report. Additionally, mitochondrial DNA can also be analyzed either using D-loop amplification protocol or with a whole mitochondrial genome protocol using Nextera® XT DNA Library Prep Kit for library preparation.

3 Reagents/Materials Required and Their Role

3.1 ForenSeq™ DNA Signature Prep Kit

This kit contains all the chemicals and reagents required for target-specific library preparation from the DNA samples simultaneously for global autosomal STRs, Y-STRs, X-STRs, and different varieties of SNPs. The kit includes DNA primer mix A which harbors primers for 58 STRs and 94 identity SNPs. However, DNA primer mix B contains primer sets for all the markers present in primer mix A and additional 56 ancestry SNPs and 22 phenotypic SNPs. Besides, the kit contains control DNA, reaction mix, enzyme mix, storage buffer, i5 and i7 indexed adaptors, other reagents, and sample purification beads.

3.2 ForenSeq™ Index Plate Fixture Kit

During cluster amplification to retain the integrity of each amplicon, indexes are added to each of them. In this process two index (i7 and i5) adapters are used to tag the DNA with a unique combination of indices. Tagging the samples is useful in later step of data analysis to distinguish among the amplified products. Index adapters and PCR2 reaction mixture are the major components of this kit which should be stored at −25 to −15 °C.

3.3 MiSeq® FGx™ Forensic Genomics System

It is the validated next-generation sequencing system to be used exclusively for forensic genomics applications. Due to its high accuracy and increased resolution, the system can analyze the degraded, low DNA samples commonly observed in forensic setup. The advantage of using this system is the simultaneous analysis of >200 genetic markers with low level of DNA input, thus increasing the efficiency of the system in comparison to the currently used CE system. The system is capable of increasing the speed and accuracy of generating sequencing results to serve as a boon to the criminal justice system.

3.4 ForenSeq™ Universal Analysis Software

It is a complete DNA to data software solution which analyzes the results originated from MiSeq FGx™ Forensic Genomics System when the samples are prepared using ForenSeq DNA Signature

Prep Kit. The software interface provides solutions for population statistics, sample comparison, as well as predicting phenotype of an individual through its ancestry, hair, and eye color estimation.

3.5 Template DNA

As sequencing is carried out during NGS, the template DNA should be prepared with utmost care to minimize the level of contamination. The optimum concentration of template DNA to be used for target-based NGS reaction is 0.2–1.0 ng/μl. Always use recommended amount of template DNA for a better result. Due to high sensitivity and specificity of the NGS reaction, prior estimation of DNA concentration by qRT-PCR is always useful. Additionally, most of the NGS library preparation kits can also use stains on FTA card directly.

4 Procedure

1. Prepare template DNA containing the recommended concentration of DNA required for a NGS reaction.
2. Use ForenSeq™ DNA Signature Prep Kit for preparing the DNA libraries using the manufacturer's guidelines, briefly as follows:
   (a) Prepare a master mixture by mixing 4.7 μl of PCR 1, 0.3 μl of FEM, and 5.0 μl of DPMA or DPMB per reaction.
   (b) Transfer 10 μl of the master mixture to each well of a 96-well plate and add 5 μl of DNA samples to each well. Additionally, add positive control and Milli-Q water as negative control to the designated wells.
   (c) Centrifuge the plate at 1000 × *g* for 30 s and run on the PCR machine at 98 °C for 3 min; followed by 8 cycles of 96 °C for 45 s, 80 °C for 30 s, 54 °C for 2 min, and 68 °C for 2 min; followed by 10 cycles of 96 °C for 30 s, 68 °C for 3 min, and finally 68 °C for 10 and 10 °C for ∞.
   (d) After completion of the PCR program, you can either store the plate at 2–8 °C for 2 days after proper sealing or proceed immediately for the next step.
3. Use ForenSeq™ Index Plate Fixture Kit for adding of adapters to the amplicons for cluster amplification using the manufacturer's protocol which is as follows:
   (a) Centrifuge the 96-well plate containing the PCR products from the previous step at 1000 × *g* for 30 s.
   (b) Arrange i7 and i5 adaptors at columns (1–12) and rows (A–H) besides the plate, respectively.
   (c) Add 4 μl of i7 and i5 adaptors to each column and row, respectively, followed by addition of 27 μl of PCR2 master mixture to each well.

(d) Centrifuge the plate at 1000 × *g* for 30 s and run on the PCR machine programmed at 98 °C for 30 s followed by 15 cycles of 98 °C for 20 s, 66 °C for 30 s, and 68 °C for 90 s followed by 68 °C for 10 min and 10 °C for ∞.

(e) After PCR program is over, you may proceed with the next step or store the plate at 2–8 °C for 7 days after proper sealing.

4. Purify the libraries from other unused/leftover primers, dNTPs, and adaptors using sample purification beads (SPB) following the manufacturer's protocol which is as follows:
   (a) Add 45 μl of sample purification beads (SPB) to each well of the 96-well plate and centrifuge the plate at 1000 × *g* for 30 s.

   (b) Add 45 μl of tagged libraries to the corresponding wells containing SPB in the 96-well plate.

   (c) Mix the contents of the plate at 1800 rpm for 2 min after proper sealing of the plate, followed by incubation at room temperature for 5 min.

   (d) Place the plate on the magnetic stand and wait for ~2 min or till the liquid becomes transparent; decant the whole supernatant from each well.

   (e) Wash the magnetic beads present on each well using 200 μl of 80% ethanol followed by incubating on magnetic stand for 30 s and removal of the supernatant.

   (f) Centrifuge the plate at 1000 × *g* for 30 s and place it on the magnetic stand.

   (g) Remove the excess ethanol from each well using a micropipette.

   (h) After removing the plate from the magnetic stand add 52.5 μl of resuspension buffer (RSB) and mix the plate contents at 1800 rpm for 2 min after proper sealing.

   (i) Incubate the plate first at room temperature followed by on magnetic stand for 2 min each.

   (j) Transfer 50 μl of the purified libraries to a fresh 96-well plate and centrifuge the plate at 1000 × *g* for 30 s for settling down of the contents.

   (k) Immediately proceed for the next step or occasionally store the plate at −25 °C till 1 year.

5. Optionally, normalize the libraries to achieve consistent cluster density.

6. Pool libraries by adding 5 μl of each library in a 1.5 ml micro-centrifuge tube, vortex, and centrifuge briefly.

7. Denature and dilute the libraries as follows:

(a) Add 36 μl of nuclease-free water, 2 μl of human sequencing control (HSC), and 2 μl of HP3 (2 N NaOH) in a microcentrifuge tube (MCT); vortex; and centrifuge briefly.

(b) Incubate the mixture at room temperature for 5 min.

(c) Mix 591 μl of hybridization buffer (HT1), 7 μl of the pooled normalized libraries, and 2 μl of HSC mixture in a 1.5 ml MCT.

(d) Vortex and centrifuge the tube briefly.

(e) Place the tube at 96 °C on a dry bath for 2 min.

(f) Mix the contents by inverting the tube multiple times and immediately snap cool the tube at −25 °C for 5 min.

8. Load the entire content on the reagent cartridge/flow cell.

9. Load flow cell by opening the "Load Flow Cell" log on the MiSeq FGx Control Software of MiSeq FGx™ instrument.

10. Load SBS solution at the designated place and make sure that the waste bottle is completely empty.

11. Load the reagent cartridge at designated place of the instrument and select "Start Run."

12. During run, monitor the progress of the run, intensities, and quality scores.

13. After completion of the run, click on the "Next" button to proceed for a post-run wash of the instrument.

14. Analyze the run results using ForenSeq™ Universal Analysis Software as follows:

    (a) Log in to the software, create a new run, import samples for analysis, and enter sample information.

    (b) Review the positive control samples (P), negative control samples (N), and quality matrices for the run (Q).

    (c) Select "Create New Analysis" button and "Generate New Analysis."

    (d) Select STR sample details table to see the allele present on the selected marker.

    (e) Select SNP sample details to see the allele present on each marker.

    (f) Select analysis and population statistics to generate random match probability, Hardy-Weinberg expectations, and linkage equilibrium.

    (g) From analysis you may select "Start Comparison" to generate the Table of Discordance.

    (h) Select "Actions" and "Phenotype estimation" from the drop-down menu.

5 Observation and Result

1. Before proceeding for result analysis, always check the positive control, negative control, and quality matrix (Fig. 31.1). In this case of amplification of positive control the HSC region can be assessed by "P," whereas reagent blank can be assessed by negative control "N" and "Q" representing the overall quality matrices for the current NGS run.
2. Outcome of the controls and quality standards can be assessed by the color indicators. Green color indicates that the sample has been sequenced with sufficient coverage and the calculated genotype is concordant. However, when green color appears, sufficient coverage may not be present for the sample and hence the genotypes are not concordant.
3. By selecting a sample for STR sample details table the following icon will appear (Fig. 31.2).
4. By clicking on a box containing the alleles, the signal intensity and nucleotide sequence can be visualized (Fig. 31.3).
5. Similarly, the SNP sample details can be visualized as follows (Fig. 31.4).

6 Precautions

- Wear gloves and goggles during the whole process of NGS.
- Always use disposable plasticware which are pre-sterilized and DNase free.
- Use filter tips to avoid cross-sample contamination.

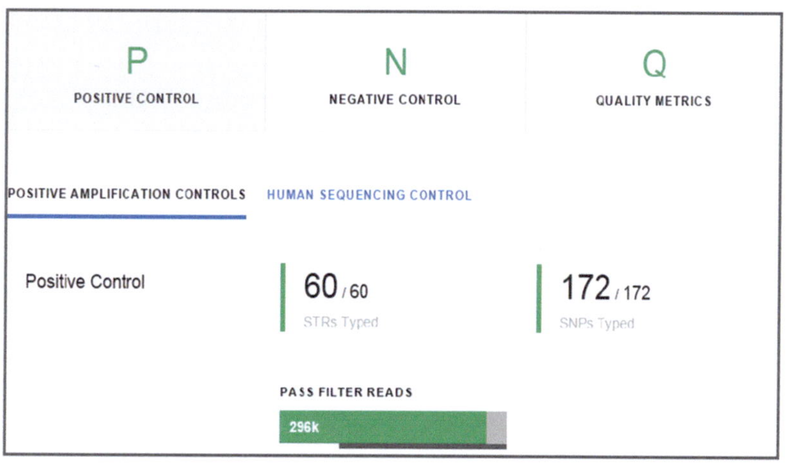

Fig. 31.1 Results of controls and quality standards of an NGS run

STRs

● Single Source Sample
● Interlocus Balance
59/59 Loci Typed

Amelogenin	D1S1656	TPOX	D2S441	D2S1338	D3S1358
X Y	17.3 18.3	8 11	11.3 14	20 23	16 18
D4S2408	FGA	D5S818	CSF1PO	D6S1043	D7S820
10 11	23 25	9 12	10 11	11 12	8 10
D8S1179	D9S1122	D10S1248	TH01	vWA	D12S391
12 14	12 12	13 15	8 8	18 18	21 21

Fig. 31.2 Table showing STR details of a sample

- Do not store the reagents under a light source.
- Always store the reagents in appropriate storage conditions.
- Never use expired reagents.
- Assess the quality and quantity of template DNA prior to setting up of an NGS reaction.
- Always prepare the master mixture for extra 10% of the samples.
- Always change tips in between samples.
- Set up PCR-1 in a pre-PCR environment.
- Avoid repeated freeze-thaw of the reagents.
- Check and remove bubbles (if any) in the PCR tubes.
- Do not forget to bring the sample purification beads to room temperature before use.
- Always change tips when adding adapters and primers for each column and row.

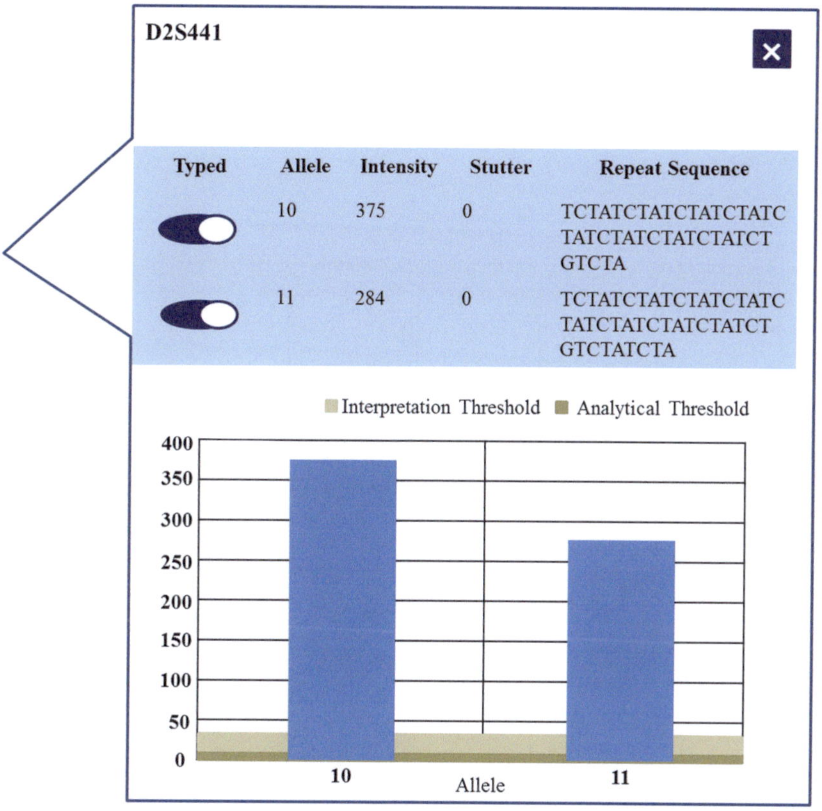

Fig. 31.3 Icon showing the signal intensity and nucleotide sequence

- Handle the reagents with utmost care as you are dealing with the hazardous known chemicals.
- Use dedicated separate waste disposal containers for solid and liquid wastes.
- Ensure that reagents in the cartridge are thawed thoroughly to enable proper sequencing.
- Never pierce any reagents on the reagent cartridge as they will be pierced automatically during the run.
- Never touch the MiSeq FGx instrument after commencement of a run as it is highly sensitive to vibration.
- Do not open the flow cell compartment during a run.
- Always perform a post-sequencing instrument wash.
- Regularly perform the maintenance wash of the sequencer ideally once in every month.

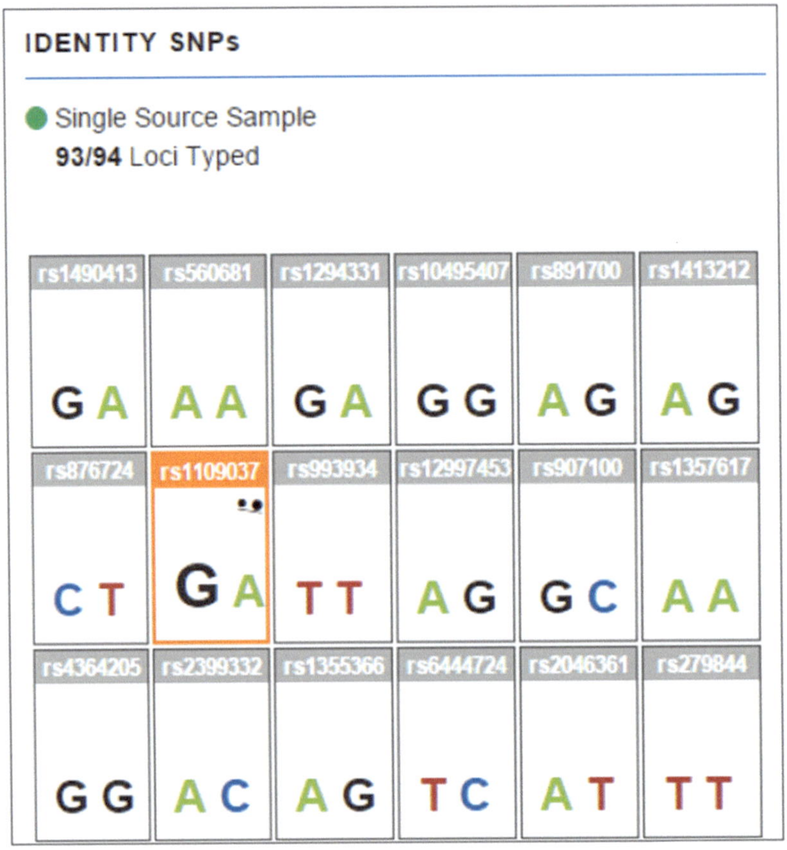

Fig. 31.4 Table showing SNP details of a sample

7 Troubleshooting

Problem	Tentative cause	Possible solutions
Low sample intensity	Less quality or quantity of samples	Reprocess the sample after re-isolating DNA
		Try with few samples for sequencing, at least eight samples in one run

(continued)

Problem	Tentative cause	Possible solutions
High cluster density	Too much library added in the mixture	Rerun the sample with sufficiently diluted libraries
	More amount of magnetic beads added	Normalize the libraries and re-sequence the samples
	Too much adapter dimer in library preparation	Check the amount of dimers in a Bioanalyzer; if the amount exceeds 5%, repeat library purification process
Low cluster density	Low/degraded DNA input	Repeat library preparation step with higher input DNA
	DNA did not denature properly	Ensure that denaturation step is performed properly
	Blockage of instrument	Perform a maintenance wash
Run incomplete	HSC not added	Add HSC and repeat the sequencing step
	Reagents not performing up to the mark	Contact technical support personnel of your instrument
Error during analysis	Analysis software is not running	Reload the software and look for the possible solution in the "help" icon corresponding to the error command
	Any other error command	Look for the possible solution in the "help" icon corresponding to the error command

Chapter 32

Forensic Trace and Touch DNA Analysis

1 Introduction

The earlier version of DNA fingerprinting analysis, i.e., restriction fragment length polymorphism (RFLP), was dependent on a huge quantity of input DNA for its successful operation. However, with the advent of polymerase chain reaction (PCR)-based short tandem repeat (STR) typing, a trace amount of DNA as low as 1 ng is required to generate a DNA profile successfully. In this regard, traces of DNA transferred to any object at the crime scene by the personnel present can be analyzed and the suspects can be linked. A healthy human sheds around tens of thousands of epithelial cells every day (Williamson 2012), which is sufficient enough to yield a DNA accounting for a successful generation of DNA profile. Additionally, many genotyping kits sensitive for DNA typing are in use nowadays favoring the chances of trace DNA analysis (Einot et al. 2017). In general, touch DNA refers to the transfer of trace amount of epithelial cells by the user/perpetrator on various objects present at the crime scene, i.e., murder weapon, vehicles, or questioned documents.

The amount of cells shed by an individual varies from person to person which is approximately 400,000 cells per day (Wickenheiser 2002). Various physiological factors also affect the shedding of cells as the nervous suspects perspire more to leave more number of cells in the crime scene (Alketbi 2018). Touching of own face or head while committing a crime increases the chance of adding more number of cells of the suspect to the crime scene. Similarly, touching a rough surface increases the chance of transfer of more number of cells than a smooth surface. Additionally, contact pressure also affects the amount of cells shed by an individual on the object which increases the probability of yielding the suspect's DNA profile.

H. R. Dash et al., *Principles and Practices of DNA Analysis: A Laboratory Manual for Forensic DNA Typing*, Springer Protocols Handbooks, https://doi.org/10.1007/978-1-0716-0274-4_32,

There are different terminologies associated with the DNA analysis with its lesser quantity of availability, i.e., trace DNA, touch DNA, low-copy DNA, and low-template DNA. However, these terminologies cannot be generalized for their use in a singlet form. Trace or touch DNA can be used for the analysis of a tiny amount of biological samples collected from the crime scene. This can be used for any sample which does not meet the recommended threshold of various stages of DNA analysis, i.e., detection, DNA extraction, PCR, and result interpretation. Low-template DNA is used to describe the process of amplification which will generate stochastic effects, whereas low copy number is generally used to describe the process of increased cycle number of PCR conditions rather than the template concentrations (van Oorschot et al. 2010). Thus, in general all touch DNA samples are trace DNA but all trace DNA samples are not touch DNA.

Thresholds of DNA amount for biological samples vary from laboratory to laboratory, i.e., 100 pg (Gill 2001) or 200 pg (Budowle et al. 2009). In addition to template threshold, all the steps involving DNA typing should be monitored properly during trace DNA analysis. It is not necessary that a trace DNA sample will remain as trace sample during all the processing steps. It is also possible that the sample may meet the threshold level during PCR process to generate a good DNA profile. Thus, all the steps of DNA analysis such as sample collection, DNA extraction, PCR, and interpretation should be carried out carefully to avoid possible contaminations.

2 Targets of Trace DNA Samples

Trace/touch DNA samples become of evidentiary value only when the possibility of innocent transfer is ruled out. The primary target surfaces carrying touch/trace DNA should be understood properly by the investigator or sample collector. In cases of murder/homicide, touch DNA samples may be targeted from the surfaces of the murder weapons such as knife, revolver, or pistol used by the perpetrator. Similarly, in sexual assault cases clothing of the victim may be used as a potential source of culprit's touch DNA samples. Additionally, certain suspected skin surfaces can be checked for the presence of any trace amount of touch DNA samples of the suspect (s). Touch DNA analysis also plays a useful role in cases of burglary where the recovered items can be analyzed for the presence of touch DNA samples of the culprits. Similarly, criminal detection of the forgerer's DNA on the forged documents can be analyzed to reach the culprit. Additionally, a wider range of exhibits can be targeted for forensic touch/trace DNA analysis, i.e., tools, clothing, vehicles, food, bedding, condoms, lip cosmetics, wallets, jewellery, glass, whole body surface, cables, windows and doors, and many others.

Beyond doubt, touch DNA plays a useful role in solving property-related issues. 8% of the success has been achieved for touch DNA analysis during 2009–2015 by the Harris County Institute of Forensic Sciences (HCIFS). A study on 252 trace DNA samples constituting 201 burglary, robbery, and drug cases shows a success rate in 14% of the cases, whereas in 21% of cases, a mixed genetic profile was obtained (Raymond et al. 2009). Despite low success rate the awareness regarding this technology is being increased among the investigators, mostly in the cases associated with burglaries and thefts. In violent criminal cases, where no blood or seminal stains are visible, the suspected articles can be submitted for touch DNA analysis generating investigative leads. Additionally, the cold cases where visible stains are not found or the samples are highly degraded can be re-examined for touch DNA analysis.

3 Collection Strategies for Trace DNA Samples

The most common technique of collecting trace/touch DNA sample is by rubbing the surface using a moistened swab. Swabbing of the target surface should be carried out perfectly by simultaneous application of pressure and rotation along the full surface area. Currently double-swabbing method is also in practice for the same purpose by the use of a moisten swab followed by a dry swab (Bartlett and Short 2017). During the process of double-swab technique, the use of moisten swab dislodges the cells from the surface, whereas the second dry swab picks the cellular material from the surface. Though water is the most commonly used moistening agent, 0.01% sodium dodecyl sulfate (SDS) and/or isopropanol is also recommended by certain studies (Prinz et al. 2006). After collection of samples in the swab it can be processed immediately or air-dried; however, air-drying reduces the DNA yield from the swab. For enhanced recovery of DNA, the swab should be kept in refrigerated conditions immediately after its collection (van Oorschot et al. 2003).

Besides the traditional methods of swabbing, tape lifting is also recommended by various workers. It can be used for lifting suspected samples from the fabrics as well as from the ridged surfaces. A study has also revealed the effectiveness of tape lifting technique in comparison to double-swabbing technique for the recovery of higher quantity of DNA (Williams et al. 2013). Besides, FTA® paper can also be used in place of common swabs. In addition to that, cutting method can be used for soft items such as clothing; however it adds unnecessary substrates into the processing tubes that increases the cost of processing per sample. Though scrapping method is also used by some laboratories using a piece of sterile scalpel blade, it is not recommended for outdoor collection of samples.

Due to the presence of small amount of DNA in trace DNA samples, they are highly prone to degradation when destructive methods are used for its detection. Hence, nondestructive methods such as use of alternative light sources and fingerprint techniques can be used for recognition of the source DNA (Sowmyya 2016). Additionally, personal protective equipment (PPE) such as gloves, masks, head, and shoe cover, and in certain instances complete body cover, can be used during collection of touch DNA samples from the crime scene.

4 Analysis of Trace DNA Samples

Trace DNA samples can be processed like other samples with extra precaution to avoid possible contaminations. The earlier DNA extraction techniques like Chelex method or manual method using phenol-chloroform reagents cannot recover around 75% of the total available DNA in a sample (Côté et al. 2008). As most of the genotyping methodologies require DNA samples in the range of 0.1–1 ng of DNA and trace DNA samples itself contain very little amount of DNA in them, any loss of DNA samples during the extraction steps can only give rise to a partial or no DNA profile. In this regard, development of a new robust DNA extraction methodology using silica-coated magnetic beads is highly useful in recovering trace amount of DNA from the substrate. Additionally, complete automation with suitable robotic systems minimizes the chance of contamination and is useful for the processing of higher number of samples in small time which is the biggest concern in processing trace DNA samples. When large volume of trace DNA sample is available it is always recommended to use the concentration devices, i.e., Microcon (Millipore), MinElute (Qiagen), or NucleoSpin (Clontech). To minimize sample loss during concentration step addition of poly A RNA or salmon sperm DNA is recommended by many users (Lever et al. 2015). Though many techniques are available for the extraction of DNA from trace DNA samples, the selection of suitable methodology is always dependent on sample, technique, and analyst's preference.

Touch DNA samples do not contain trace amount of DNA always. Hence, it is recommended to quantify these samples prior to their further processing which gives an indication of the tentative DNA profile to be generated. Out of many techniques available for DNA quantitation, probe-based qRT-PCR technique is considered to be the gold standard for quantification of touch DNA extract. The commercially available kits for DNA quantification have been listed in Appendix B. The next step for analysis involves the multiplex PCR by various genotyping kits. As the sensitivity of most of the genotyping kits falls in the range of 0.1–1 ng, the presence of DNA in such quantity generates a complete DNA profile. Otherwise a partial DNA profile is obtained.

Technological advancement has revolutionized the analysis of touch DNA samples where importance has been given to avoid allele drop in/out in DNA profiles. The new generation kits such as AmpFlSTR®, Identifiler® Direct, GlobalFiler™ Express, and PowerPlex® 18D have the capability of direct PCR from the samples minimizing allele dropout rate. Additionally, the mini-STR analysis can also be performed to generate a complete analyzable profile from the low-template touch DNA samples (Nieuwerburgh et al. 2014). Currently, massive parallel sequencing of SNP markers for ancestry and phenotype information is being used which requires a little amount of DNA compared to the STR-based capillary electrophoresis, and thus it becomes highly useful for touch DNA analysis (Bruijns et al. 2018). In this regard, some researchers have also tried with the addition of extra Taq Polymerase (Kloosterman and Kersbergen 2003) and increase in the PCR cycle conditions (Whitaker et al. 2001) to get an improved DNA profile from touch DNA samples.

5 Concerns and Limitations of Trace DNA Analysis

The most controversial issue pertaining to trace DNA analysis is the result interpretation. Though detailed guidelines are in existence for the interpretation of trace DNA result, its implementation throughout the global laboratories is in question till now which raises a huge concern for its application in the medicolegal cases. Preferential amplification leading to allele dropout is a huge problem for low DNA samples. Additionally, confirming the zygosity of a trace DNA sample becomes extremely difficult due to high stochastic effect and peak height imbalance. Drop-in of alleles also creates a huge problem in such types of samples creating confusion regarding the number of contributors in a STR profile (Hong 2014).

Interpretation of mixed profile is still in debate and it becomes extremely difficult for the profiles from low DNA samples due to a higher chance of allele drop-in, dropout, and stutter percentage. Many software systems are available nowadays to deal with this problem and separate the contributor's profiles in a mixed profile depending on the quantitative models of peak heights and likelihood ratios. Some of the software programs include TrueAllele, STRmix, DNA VIEW Mixture Solution™, LiRaHT, DNAmixtures, EuroForMix, likeLTD, CEESIt, and Kongoh (Manabe et al. 2017). These mixed DNA profiles in touch DNA samples may arise from the background contaminations and leftover DNA present at the crime scene. These left over DNA samples do not create a significant problem in normal cases as these minute peaks are masked in the large quantity of target DNA due to preferential amplification. These contaminant DNA may be deposited either before the crime, during committing the crime, during crime investigation, or during experimentation procedure. Though contamination generated prior to the crime and during the crime cannot be controlled, it can be minimized during sample collection and their processing steps by adopting a standard operating procedure (Llamas et al. 2017).

Though many studies have been conducted so far to increase the efficacy of touch DNA analysis, the technique remains unavailable till date to distinguish among the origin of DNA due to primary, secondary, or tertiary transfer. The sequence of events during primary, secondary, and tertiary transfer of DNA is illustrated in Fig. 32.1. In primary transfer, the DNA deposits directly on an object when an individual touches it. Transfer of DNA on an object through another individual giving rise to a mixed DNA profile is called as secondary transfer (Edwards et al. 1991). Similarly, tertiary transfer of DNA can also occur with the involvement of third party leading to its DNA deposition on the object. In such cases, a mixed DNA profile from three sources is observed on the object. In this regard, it becomes imperative to distinguish between these three types of DNA transfers and unfold the chain of

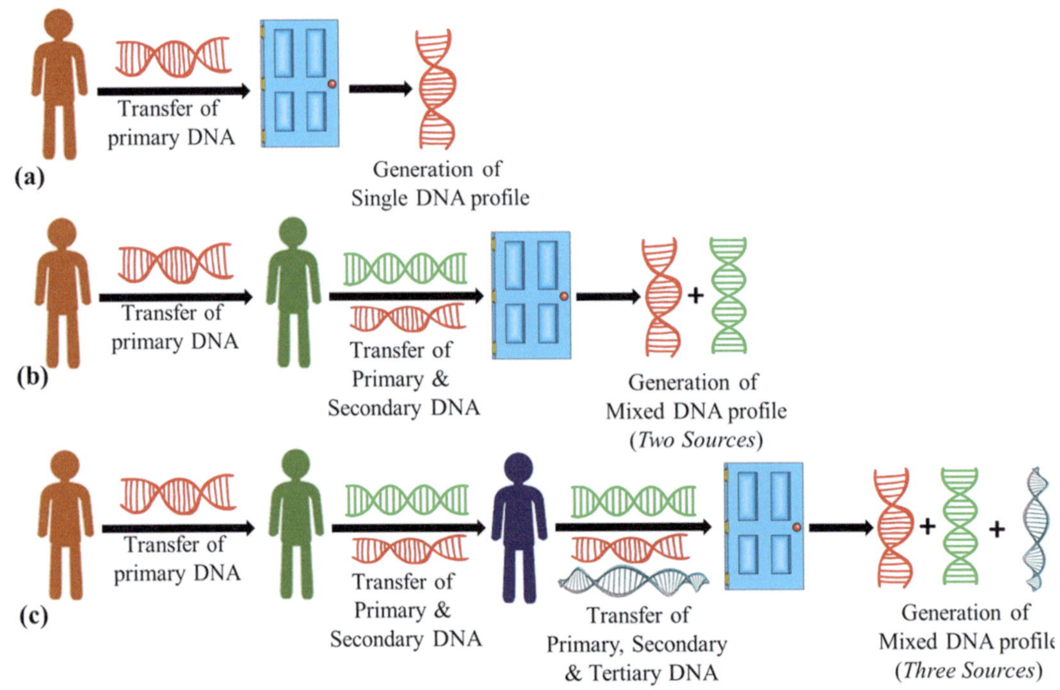

Fig. 32.1 Mechanism of (**a**) primary, (**b**) secondary, and (**c**) tertiary touch DNA transfers

incidences that lead to the generation of mixed DNA profiles on an object when touch DNA samples are analyzed (Goray et al. 2010).

6 Conclusion

Forensic trace or touch DNA analysis is of huge importance in criminal justice system. However, till date no gold standard technique is available to deal with such extremely low-template DNA samples. Over the years trace DNA analysis has been improved with the introduction of magnetic bead-based DNA extraction and robust, highly sensitive kits for STR amplification. However, other problems associated with trace DNA analysis are still persisting such as contamination issues, mixture interpretation, and resolving among primary, secondary, and tertiary transfer of DNA. Thus, proper training, methodology development, policy making, and proficiency testing will increase the usefulness of trace DNA typing in forensic investigations.

Chapter 33

RAPID DNA Technology: A Boon to Forensic DNA Typing

1 Introduction

With the emergence of forensic DNA typing technology, criminal justice system has considered it to be the gold standard among other evidences. During the processing of forensic samples, there are many labor-intensive processes such as DNA isolation, quantification, amplification of short tandem repeat (STR) markers by multiplex PCR, genotyping by capillary electrophoresis, and data interpretation. Traditional forensic DNA typing technology requires a number of expensive instruments and trained manpower. Routine samples may take at least a day per sample for analysis. Considering the backlogs and waiting period, which is a common fact for most of the DNA typing laboratories, quite a long time is required for obtaining a DNA fingerprinting result. With the increase in demand from the law enforcement agencies, time-bound result from the demanded samples has become the need of the hour. In this context, the recently discovered completely automated platforms are on demand for the forensic DNA typing laboratories.

The RAPID DNA instruments provide completely automatic solutions from sample to results without any human intervention. These commercially available robotic systems have integrated all the requisite steps of forensic DNA typing in a single machine to minimize the processing time as well as the chance of contamination during manual handling of samples. Besides shortening the processing time, ease of operating this instrument has also advocated its use in the field conditions.

H. R. Dash et al., *Principles and Practices of DNA Analysis: A Laboratory Manual for Forensic DNA Typing*, Springer Protocols Handbooks, https://doi.org/10.1007/978-1-0716-0274-4_33,

2 The Technology

The technology uses a prefilled cartridge containing all the chemicals required for the routine processes of DNA fingerprinting. After insertion of the sample to the instrument, the operation commences with automatic calibration. In this regard, it is useful for the semiskilled personnel in the field to generate a DNA profile. Due to low processing time, the machine allows identification of criminal(s) within 75–90 min and is helpful in exonerating the suspect(s) in a time-bound manner. The principle behind the operation of this instrument lies with the micro-fluidic processing technique. Micro-fluidic processing technique utilizes smaller amounts of chemicals, hence minimizing the operation cost, reaction time, as well as size of the instrument and the cartridges containing chemicals. All the steps including extraction, amplification, electrophoresis, and data interpretation take place in one micro-fluidic chip.

A typical RAPID DNA instrument consists of three parts, i.e., the swab, a chip, and the instrument. The swab is required for the input of desired amount of biological samples to the instrument for analysis. The chip contains all the reagents, materials, and waste containments necessary for the process which include reagents for DNA extraction, amplification by PCR, and separation of amplified fragments by capillary electrophoresis. The samples present in the swab are taken by pneumatic pressure and at no stage there is the direct contact between the instrument and the sample or reagents. This limits the chance of possible cross contamination. The additional advantage of this instrument is that the generated DNA profiles are compatible to the CODIS, Expanded European Standard Set, Australia's National Criminal Investigation DNA Database, Canada's National DNA Data Bank, China's National DNA Database, Germany's DNA-Analyze-Datei, New Zealand's National DNA profile databank, and United Kingdom's National DNA Database (Carney et al. 2019). The various steps performed by a RAPID DNA system are described in Fig. 33.1.

Cases of various natures are continuously being solved promptly by using RAPID DNA technology such as reuniting of families separated during mass migration or natural calamities, prevention of human trafficking, and identification of mass casualty victims. Though the RAPID DNA technology has received huge applause throughout the globe, it faces many challenges courtesy variant nature of the samples pertaining to forensic cases. Some of the challenges of RAPID DNA technology are as follows:

1. All sample types mostly the tough samples, i.e., bone and teeth, cannot be processed using the currently available RAPID DNA technology.

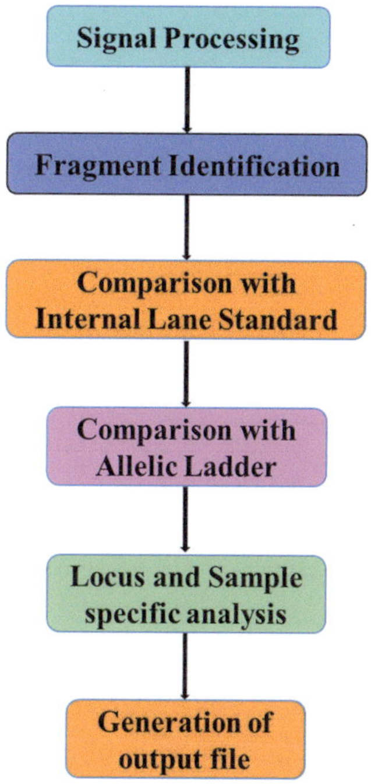

Fig. 33.1 Various steps performed by the RAPID DNA system

2. Some samples also require pre-processing limiting the sample to result nature of the technology.
3. The technology is yet to reach a ~100% result in all the samples tested.
4. Non-cost-effectiveness in comparison to the traditional DNA technology for varied nature of the forensic case samples.
5. Pendency and backlog of cases may result in degraded samples, where the current technology is of limited applications.

3 Currently Available Technologies

The first fully automatic "swab in-profile out" RAPID DNA analysis approved by FBI is the NetBio (http://netbio.com/) DNAscan Rapid DNA Analysis™ (Network Biosystems). The two systems of NetBio, i.e., ANDE™ and DNAscan™, use a common cotton swab for input of reference samples to be processed within 90 min by the instruments. The currently used DNAscan 6C has the capability to analyze four samples at 27 loci within 90 min.

Another RAPID DNA instrument in the market is Rapid-HIT™ 200 system (IntergenX) which is slightly smaller in size in comparison to NetBio instruments and is capable of processing up to eight samples at the same time. The second-generation

IntergenX (https://integenx.com/) instrument RapidHIT ID has reduced its cost to half of the first-generation instrument and is capable of using both GlobalFiler Express and NGMSelect 17 kit.

Another fully automatic portable DNA Analyzer has been manufactured by NEC (http://in.nec.com/en_IN/products/public-safetysecurity/product/portable-dna-analyzer.html). This instrument is highly popular in the Asian subcontinent and is capable of performing the entire DNA analysis within 25 min. As this instrument is portable enough to be carried to the crime scene, it is highly useful in speeding up of the crime investigation and prevention processes. Recently LGC (https://www.lgcgroup.com/products/paradnatechnology/#.WJ4-KcLHxD9) has manufactured ParaDNA® instruments capable of providing high success rate, fastness, and reliable profile generation in the quickest possible time.

Currently, National DNA Index System (NDIS) is the responsible agency to provide approval of the RAPID DNA systems for forensic applications. As the process has limitation for analyzing the low-DNA-content (LDC) samples, many developmental validations are in progress for improving the associated methodologies. Incorporation of single-use BioChipSet (BCS) has increased the performance of the technology to generate a readable DNA profile from as low as 5 pg/μl of DNA samples (Turingan et al. 2016; Buscaino et al. 2018).

4 Conclusion

RAPID DNA technology possesses a definitive advantage for its use in definitive scenario such as at the crime scene. Due to ease in operation and complete automation, processing of the samples at crime scene or police stations may not require highly skilled personnel besides interpretation of the obtained results. Regarding cost of the instrument and its associated consumables, if we include the operational as well as infrastructure cost, the RAPID DNA technology is highly useful for processing fewer samples with the requirement of urgent results for investigation purpose. Though the technology has improved a lot since its inception, it is still not so useful in challenging forensic samples. Hence, the technology works fine for database generation and analysis of reference samples in a short period of time.

Appendix A: Composition of Reagents, Buffers, and Solutions

Lysis Buffer I

Tris: 30 mM

EDTA: 5 mM

NaCl: 50 mM

Lysis Buffer II

NaCl: 75 mM

EDTA: 2 mM

Forensic Buffer (pH 8.2)

Tris: 100 mM

EDTA: 5 mM

NaCl: 50 mM

Hair Digestion Buffer

Mix the following stock solution as per the following:

1 M Tris–HCl (pH 7.5): 1 ml

0.5 M EDTA: 2 ml

5 M NaCl: 10 ml

20% SDS: 10 ml

H. R. Dash et al., *Principles and Practices of DNA Analysis: A Laboratory Manual for Forensic DNA Typing*, Springer Protocols Handbooks, https://doi.org/10.1007/978-1-0716-0274-4, © Springer Science+Business Media, LLC, part of Springer Nature 2020

Deionized water: 86 ml

Note: Do not autoclave, and store at 4 °C till further use.

Proteinase-K (20 mg/ml)

Add 20 mg of proteinase-K in 500 μl of TE buffer first and then add 500 μl of 100% glycerol. Mix well with slow pipetting, prepare aliquots, and store at −20 °C.

Note: As the protein is soluble in water, the solvent (water/buffer) to be used depends on the condition of the supplied powder. If proteinase-K powder is supplied in a lyophilized form using a buffer, it can be dissolved in deionized water. However, if it is lyophilized from water, it should be dissolved in TE buffer. Thus, check the manufacturer's guidelines before dissolving proteinase-K.

1 M DTT

Weigh out 3.86 g of DL-dithiothreitol (DTT) ($C_4H_{10}O_2S_2$) of molecular weight 154.25 in a 50 ml disposable flask. Add 20 ml of deionized water to it and mix well to dissolve the DTT completely. Make up the volume to 25 ml with deionized water. Sterilize the solution by passing through 0.22 μm syringe filter under aseptic conditions. Store the solution at −20 °C till further use.

Caution: Do not autoclave the DTT solution; autoclaving destroys the DTT.

20% (w/v) SDS

Sodium dodecyl sulfate (SDS) has the molecular formula of $CH_3(CH_2)_{11}OSO_3Na$. To prepare 100 ml of 20% SDS solution, take 20 g of SDS in a 250 ml of conical flask and add 80 ml of deionized water to it. Mix gently using a magnetic stirrer. Optionally you may heat the solution at 68 °C using a water bath for thorough dissolving. Adjust the volume to 100 ml using deionized water. Make small aliquots and store at room temperature.

Note: Autoclaving is not required for SDS solution. Wear mask while preparing SDS solution to avoid inhalation of the fine powder.

Tris-Saturated Phenol (TSP)

Open 100 g of phenol in a fume hood and pour in a 100 ml of 50 mM Tris-Cl pH 8.0. Close the lid of the bottle and mix gently.

Allow it to stand for 1–2 h till phenol liquefies and two phases get separated. Remove the supernatant using a glass pipette and further add 100 ml of 50 mM Tris-Cl pH 8.0. Close the bottle and mix gently. Allow it to stand for 1–2 h till phenol liquefies and two phases get separated. Repeat the saturation step 4–5 times till pH of the supernatant reaches 7–8. Dispense in aliquots and add 10 ml of 50 mM Tris-Cl (pH 8.0) to each aliquot. Store at −20 °C till further use.

Alternatively you may try using the commercially available Tris-saturated phenol (TSP). Some manufacturers add stabilizers in the solution. As pure phenol is colorless only oxidation in phenol changes its color to pale yellow to pink. Storing the solution at −20 °C minimizes the risk of oxidation of the solution.

Chloroform: Isoamyl Alcohol (24:1)

Mix 96 ml of chloroform with 4 ml of isoamyl alcohol (3-methyl-1-butanol) in a glass measuring cylinder. Make small aliquots and store at room temperature.

3 M Sodium Acetate

Weigh 24.61 g of sodium acetate (CH_3COONa) and transfer in a 250 ml of conical flask. Add 80 ml of deionized water and mix until sodium acetate is completely dissolved. Adjust the pH to 5.2 by using glacial acetic acid. Finally, make up the volume to 100 ml with deionized water. Sterilize the solution by autoclaving at 121 °C and 15 pounds per square inch pressure for 15 min. Allow the solution to cool. Dispense in aliquots and use.

70% Ethanol

Use the formula of $N1 \times V1 = N2 \times V2$ to prepare the desired quantity of 70% ethanol solution from the absolute ethanol (95% solution). For example, to prepare 200 ml of 70% ethanol solution, dissolve 147.4 ml of 95% ethanol with 52.6 ml of deionized water. Mix well before use.

Tris-EDTA (TE) Buffer (pH 8.0)

The constituents of TE buffer are 10 mM Tris [Tris (hydroxylmethyl) aminomethane] and 0.5 M EDTA [diaminoethane tetra-acetic acid] (pH 8.0). Prepare 1 M stock of Tris by dissolving 60.57 g of Tris in 500 ml of deionized water after adjusting the pH to 8.0. Prepare 0.5 M EDTA (pH 8.0) solution by dissolving 18.6 g of EDTA in 100 ml of deionized water and adjusting the pH to 8.0. To prepare 500 ml of TE buffer, mix 5 ml of 1 M Tris (pH 8.0) and 1 ml of 0.5 M EDTA (pH 8.0). Make up the volume to 500 ml by adding 496 ml of deionized water. Sterilize the solution by autoclaving at 121 °C and 15 pounds per square inch pressure for 15 min.

Caution: Disodium EDTA salt will not dissolve in the solution until pH 8.0 is achieved.

Normal Saline Solution (0.85%)

Dissolve 8.5 g of NaCl in deionized water. Sterilize the solution by autoclaving at 121 °C and 15 pounds per square inch pressure for 15 min. Allow the solution to cool. Dispense in aliquots and use.

0.5 M EDTA (pH 8.0)

Add 186.1 g of disodium EDTA.2H_2O salt to 800 ml of deionized water. Mix vigorously on a magnetic stirrer. Add NaOH pellets to adjust the pH to 8.0. Make up the volume to 1000 ml using deionized water. Sterilize the solution by autoclaving at 121 °C and 15 pounds per square inch pressure for 15 min. Allow the solution to cool. Dispense in aliquots and use.

Caution: Disodium EDTA salt will not dissolve in the solution until pH 8.0 is achieved.

Appendix B: List of Commercially Available Kits and Instruments for Forensic DNA Analysis

(a) List of Commercially Available DNA Extraction Kits/Systems

Sl. no.	Name of kits	Manufacturer's name and address
1	QIAamp Micro DNA kit	Qiagen, Valencia, CA
2	QIAamp DNA Investigator Kit	
3	EZ1 DNA Investigator Kit	
4	QIAsymphony DNA Investigator Kit	
5	DNA IQ™ System and DNA IQ Reference Sample Kit for Maxwell® 16	Promega Corporation, Madison, WI
6	DNA IQ Casework Sample Kit for Maxwell 16	
7	PrepFiler™ Forensic DNA Extraction Kit	Thermo Fisher, South San Francisco, CA
8	PrepFiler Express™ DNA Extraction Kit	
9	Forensic GEM Kits	ZyGEM, Hamilton, New Zealand
10	MagPurix Forensic DNA Extraction Kit	Zinexts, New Taipei City, Taiwan
11	MPure™ Forensic DNA Extraction Kit	MP Biomedicals, Santa Ana, CA
12	E.Z.N.A. Forensic DNA Kit	Omega Bio-tek, Inc., NorCross, GA

(continued)

H. R. Dash et al., *Principles and Practices of DNA Analysis: A Laboratory Manual for Forensic DNA Typing*, Springer Protocols Handbooks, https://doi.org/10.1007/978-1-0716-0274-4,

Sl. no.	Name of kits	Manufacturer's name and address
13	Forensic DNA Extraction Kit	GeneON GmbH, Deutschland/Germany
14	GF-1 Forensic DNA Extraction Kit	Vivantis Technologies, Selangor Darul Ehsan, Malaysia
15	MagListo™ 5M Forensic Sample DNA Extraction Kit	Vivantis Technologies, Selangor Darul Ehsan, Malaysia
16	AGOWA® mag Maxi DNA Isolation Kit	AGOWA, Germany
17	InviMag Forensic Kit	Invitek mbH, Germany
18	UltraClean™ Forensic DNA Isolation Kit	Mo Bio Laboratories Inc., Carlsbad, CA
19	HiPurA™ Forensic Sample Genomic DNA Purification Kit	HiMedia Laboratories, Mumbai, India
20	HiPurA™ Quick Bone DNA purification kit	HiMedia Laboratories, Mumbai, India

(b) List of Commercially Available DNA Quantitation Kits/Systems

Sl. no.	Name of kits	Manufacturer's name and address
1	Quantifiler Trio	Thermo Fisher, South San Francisco, CA
2	Quantifiler HP	
3	Quantifiler Duo	
4	Quantifiler Human	
	Quantifiler® Y human male DNA quantification kit	
5	Investigator® Quantiplex HYres Kit	Qiagen, Valencia, CA
	Investigator® Quantiplex Pro Kit	
	Investigator® Quantiplex Kit	
6	PowerQuant® System	Promega Corporation, Madison, WI
	Plexor® HY System	
7	InnoQuant™ Human DNA Quantification & Degradation Assessment Kit	InnoGenomics Technologies, LLC

(c) List of Commercially Available Genotyping Kits

Sl. no.	Name of the kit	List of STR markers	Manufacturer's name and address
1	AmpF/STR® Identifiler® kit	D8S1179, D21S11, D7S820, CSF1PO, D3S1358, TH01, D13S317, D16S539, D2S1338, D19S433, vWA, TPOX, D18S51, amelogenin, D5S818, FGA	Thermo Fisher Scientific, Foster City, CA, USA
2	AmpF/STR® Identifiler® Plus kit	CSF1PO, D2S1338, D3S1358, D5S818, D7S820, D8S1179, D13S317, D16S539, D18S51, D19S433, D21S11, FGA, TH01, TPOX, vWA, amelogenin	Thermo Fisher Scientific, Foster City, CA, USA
3	AmpF/STR® Identifiler® Direct kit	D8S1179, D21S11, D7S820, CSF1PO, D3S1358, TH01, D13S317, D16S539, D2S1338, D19S433, vWA, TPOX, D18S51, D5S818, FGA, amelogenin	Thermo Fisher Scientific, Foster City, CA, USA
4	AmpF/STR® MiniFiler™ kit	CSF1PO, FGA, D16S539, D18S51, amelogenin, D2S1338, D21S11, D7S820	Thermo Fisher Scientific, Foster City, CA, USA
5	AmpFℓSTR® NGM™ kit	D3S1358, vWA, D16S539, D2S1338, D8S1179, D21S11, D18S51, D19S433, TH01, FGA, D1S1656, D12S391, D10S1248, D22S1045, D2S441, amelogenin	Thermo Fisher Scientific, Foster City, CA, USA
6	AmpFℓSTR® NGM SElect™ kit	D3S1358, vWA, D16S539, D2S1338, D8S1179, D21S11, D18S51, D19S433, TH01, FGA, amelogenin, D10S1248,	Thermo Fisher Scientific, Foster City, CA, USA

(continued)

Sl. no.	Name of the kit	List of STR markers	Manufacturer's name and address
		D22S1045, D2S441, D1S1656, D12S391, SE33	
7	AmpFℓSTR® NGM SElect™ Express	D3S1358, vWA, D16S539, D2S1338, D8S1179, D21S11, D18S51, D19S433, TH01, FGA, amelogenin, D10S1248, D22S1045, D2S441, D1S1656, D12S391, SE33	Thermo Fisher Scientific, Foster City, CA, USA
8	AmpFℓSTR® NGM Detect™ kit	FGA, TH01, vWA, D3S1358, D8S1179, D18S51, D21S11, D12S391, D1S1656, D2S441, D10S1248, D22S1045, D16S539, D2S1338, D19S433, SE33, IQCS, IQCL, Y indel, amelogenin	Thermo Fisher Scientific, Foster City, CA, USA
9	GlobalFiler™ kit	D3S1358, vWA, D16S539, CSF1PO, TPOX, D8S1179, D21S11, D18S51, D2S441, D19S433, TH01, FGA, D22S1045, D5S818, D13S317, D7S820, SE33, D10S1248, D1S1656, D12S391, D2S1338, amelogenin, DYS391, Y indel	Thermo Fisher Scientific, Foster City, CA, USA
10	GlobalFiler™ Express kit	D3S1358, vWA, D16S539, CSF1PO, TPOX, D8S1179, D21S11, D18S51, D2S441, D19S433, TH01, FGA, D22S1045, D5S818, D13S317, D7S820, SE33, D10S1248, D1S1656, D12S391, D2S1338, amelogenin, DYS391, Y indel	Thermo Fisher Scientific, Foster City, CA, USA

(continued)

Sl. no.	Name of the kit	List of STR markers	Manufacturer's name and address
11	VeriFiler™ Express kit	D3S1358, vWA, D16S539, CSF1PO, TPOX, D8S1179, D21S11, D18S51, D2S441, D19S433, TH01, FGA, D22S1045, D5S818, D13S317, D7S820, D10S1248, D1S1656, D12S391, D2S1338, D6S1043, Penta D, Penta E, Y indel, amelogenin	Thermo Fisher Scientific, Foster City, CA, USA
12	AmpFℓSTR® Yfiler® kit	DYS19, DYS385a/b, DYS389I/II, DYS390, DYS391, DYS392, DYS393, DYS438, DYS439, DYS437, DYS448, DYS456, DYS458, DYS635 (YGATAC4), YGATAH4	Thermo Fisher Scientific, Foster City, CA, USA
13	AmpFℓSTR® Yfiler® Direct kit	DYS19, DYS385a/b, DYS389I/II, DYS390, DYS391, DYS392, DYS393, DYS438, DYS439, DYS437, DYS448, DYS456, DYS458, DYS635 (YGATAC4), YGATAH4	Thermo Fisher Scientific, Foster City, CA, USA
14	Yfiler™ Plus kit	DYS576, DYS389I, DYS635, DYS389II, DYS627, DYS460, DYS458, DYS19, YGATAH4, DYS448, DYS391, DYS456, DYS390, DYS438, DYS392, DYS518, DYS570, DYS437, DYS385a/b, DYS449, DYS393, DYS439, DYS481, DYS387S1, DYS533	Thermo Fisher Scientific, Foster City, CA, USA
15	PowerPlex® 18D System	D18S51, D21S11, TH01, D3S1358, FGA, TPOX, D8S1179,	Promega Corporation, Madison, WI

(continued)

Sl. no.	Name of the kit	List of STR markers	Manufacturer's name and address
		vWA, CSF1PO, D16S539, D7S820, D13S317, D5S818, amelogenin, Penta E, Penta D, D2S1338, D19S433	
16	PowerPlex® 21 System	Amelogenin, D1S1656, D2S1338, D3S1358, D5S818, D6S1043, D7S820, D8S1179, D12S391, D13S317, D16S539, D18S51, D19S433, D21S11, amelogenin, CSF1PO, FGA, Penta D, Penta E, TH01, TPOX, vWA	Promega Corporation, Madison, WI
17	PowerPlex® CS7 System	LPL, F13B, FESFPS, F13A01, Penta D, Penta C, Penta E	Promega Corporation, Madison, WI
18	PowerPlex® ESX 16 and ESI 16 Fast Systems	D18S51, D21S11, TH01, D3S1358, amelogenin, D16S539, D2S1338, D1S1656, D10S1248, FGA, D8S1179, vWA, D22S1045, D19S433, D12S391, D2S441	Promega Corporation, Madison, WI
19	PowerPlex® ESX 17 and ESI 17 Fast Systems	D3S1358, D8S1179, D18S51, D21S11, FGA, TH01, vWA, D2S441, D10S1248, D22S1045, D1S1656, D12S391, D2S1338, D16S539, D19S433, SE33, amelogenin	Promega Corporation, Madison, WI
20	PowerPlex® Fusion 6C System	CSF1PO, FGA, TH01, vWA, D1S1656, D2S1338, D2S441, D3S1358, D5S818, D7S820, D8S1179, D10S1248, D12S391, D13S317, D16S539, D18S51, D19S433, D21S11, amelogenin, DYS391, Penta D, Penta E, D22S1045, TPOX, SE33, DYS570, DYS576	Promega Corporation, Madison, WI

(continued)

Sl. no.	Name of the kit	List of STR markers	Manufacturer's name and address
21	PowerPlex® Fusion System	CSF1PO, FGA, TH01, TPOX, vWA, D3S1358, D5S818, D7S820, D8S1179, D13S317, D16S539, D18S51, D21S11, D10S1248, D22S1045, D2S441, D1S1656, D12S391, amelogenin, DYS391, Penta D, Penta E, D2S1338, D19S433	Promega Corporation, Madison, WI
22	PowerPlex® Y23 System	DYS576, DYS389I/II, DYS448, DYS19, DYS391, DYS481, DYS549, DYS533, DYS438 (penta), DYS437, DYS570, DYS635, DYS390, DYS439, DYS392, DYS643 (penta), DYS393, DYS458, DYS385a/b, DYS456, Y-GATA-H4	Promega Corporation, Madison, WI
23	Investigator 24plex QS Kit	Amelogenin, TH01, D3S1358, vWA, D21S11, TPOX, DYS391, D1S1656, D12S391, SE33, D10S1248, D22S1045, D19S433, D8S1179, D2S1338, D2S441, D18S51, FGA, QS1, D16S539, CSF1PO, D13S317, D5S818, D7S820, QS2	QIAGEN, Hilden, Germany
24	Investigator 24plex GO! Kit	Amelogenin, TH01, D3S1358, vWA, D21S11, TPOX, DYS391, D1S1656, D12S391, SE33, D10S1248, D22S1045, D19S433, D8S1179, D2S1338, D2S441, D18S51, FGA, QS1, D16S539, CSF1PO, D13S317, D5S818, D7S820, QS2	QIAGEN, Hilden, Germany

(continued)

Sl. no.	Name of the kit	List of STR markers	Manufacturer's name and address
25	Investigator ESSplex SE QS Kit	QS1, amelogenin, TH01, D3S1358, vWA, D21S11, QS2, D163539, D1S1656, D19S433, SE33, D10S1248, D22S1045, D12S391, D8S1179, D2S1338, D2S441, D18S51, FGA	QIAGEN, Hilden, Germany
26	Investigator ESSplex SE GO! Kit	Amelogenin, TH01, D3S1358, vWA, D21S11, D16S539, D1S1656, D19S433, SE33, D10S1248, D22S1045, D12S391, D8S1179, FGA, D2S1338, D2S441, D18S51	QIAGEN, Hilden, Germany
27	Investigator ESSplex SE Plus Kit	Amelogenin, TH01, D3S1358, vWA, D21S11, D16S539, D1S1656, D19S433, D8S1179, D2S1338, D10S1248, D22S1045, D12S391, FGA, D2S441, D18S51	QIAGEN, Hilden, Germany
28	Investigator ESSplex Plus Kit	Amelogenin, TH01, D3S1358, vWA, D21S11, D16S539, D1S1656, D19S433, D8S1179, D2S1338, D10S1248, D22S1045, D12S391, FGA, D2S441, D18S51	QIAGEN, Hilden, Germany
29	Investigator IDplex Plus Kit	Amelogenin, TH01, D3S1358, vWA, D21S11, TPOX, D7S820, D19S433, D5S818, D2S1338, D16S539, CSF1PO, D13S317, FGA, D18S51, D8S1179	QIAGEN, Hilden, Germany

(continued)

Sl. no.	Name of the kit	List of STR markers	Manufacturer's name and address
30	Investigator IDplex GO! Kit	Amelogenin, TH01, D3S1358, vWA, D21S11, TPOX, D7S820, D19S433, D5S818, D2S1338, D16S539, CSF1PO, D13S317, FGA, D18S51, D8S1179	QIAGEN, Hilden, Germany
31	Investigator Argus X-12 QS Kit	QS1, amelogenin, DXS10103, DXS8378, DXS10101, DXS10134, DXS10074, DXS7132, DXS10135, DXS7423, DXS10146, DXS10079, DXSHPRTB, DXS10148, D21S11	QIAGEN, Hilden, Germany
32	Investigator HDplex Kit	Amelogenin, D7S1517, D3S1744, D12S391, D2S1360, D6S474, D4S2366, D8S1132, D5S2500, D18S51, D21S2055, D10S2325, SE33	QIAGEN, Hilden, Germany
33	Investigator Argus Y-12 QS Kit	DYS439, DYS437, DYS390, DYS385, DYS391, DYS389-I, DYS19, DYS389-II, DYS393, DYS438, DYS392	QIAGEN, Hilden, Germany
34	COrDIS Plus kit	D3S1358, D5S818, D7S820, D8S1179, D13S317, D16S539, D18S51, D21S11, CSF1PO, FGA, TH01, TPOX, VWA, D1S1656, D2S441, D10S1248, D12S391, D22S1045 and SE33, amelogenin	Gordiz, Moskva, Russia
35	COrDYS-Y kit	DYS19, DYS389I/II, DYS390, DYS391, DYS392, DYS393, DYS385a/b, DYS438 and DYS439, DYS437, DYS456, DYS635, DYS448, DYS576, DYS481, DYS449	Gordiz, Moskva, Russia

(continued)

Sl. no.	Name of the kit	List of STR markers	Manufacturer's name and address
36	COrDX kit	DXS10148, DXS10135, DXS8378, DXS7132, DXS10079, DXS10075, DXS10074, DXS10103, HPRTB, DXS10101, DXS10146, DXS8377, DXS10134, DXS7423	Gordiz, Moskva, Russia
37	iPLEX-STR™ Kit	Amelogenin, D3S1358, TH01, D12S391, D1S1656, D10S1248, D2S441, D7S820, D13S317, FGA, TPOX, D18S51, D16S539, D8S1179, CSF1PO, D5S818, vWA, D21S11, SE33	Independent Forensics, 500 Waters Edge Lane, Lombard
38	iPLEX-STR™ Y Kit	DYS19, DYS389I, DYS389II, DYS390, DYS391, DYS392, DYS393, DYS385a/b, DYS438, DYS439, DYS437, DYS456, DYS635, DYS448, DYS576, DYS481, DYS449	Independent Forensics, 500 Waters Edge Lane, Lombard
39	iPLEX X Kit	DXS10148, DXS10135, DXS8378, DXS7132, DXS10079, DXS10075, DXS10074, DXS10103, HPRTB, DXS10101, DXS10146, DXS8377, DXS10134, DXS7423	Independent Forensics, 500 Waters Edge Lane, Lombard
40	SureID® 23comp Human DNA Identification Kit	Amelogenin, D18S1364, D1S1656, D13S325, D9S1122, D4S2366, D3S1744, D12S391, D11S2368, D21S2055, D20S482, D8S1132, D7S3048, D2S441, D19S253, D10S1248, D17S1301, D22-GATA198B05, D16S539, D6S474, D14S1434, D15S659	HEALTH Gene Technologies Co. Ltd., China

(continued)

Sl. no.	Name of the kit	List of STR markers	Manufacturer's name and address
41	SureID® 21G Human STR Identification Kit	CSF1PO, FGA, TH01, TPOX, vWA, D3S1358, D5S818, D7S820, D8S1179, D13S317, D16S539, D18S51, D21S11, D12S391, D19S433, D1S1656, D2S1338, D6S1043, Penta E, Penta D, Amelogenin	HEALTH Gene Technologies Co. Ltd., China
42	SureID® 27Y Human STR Identification Kit	DYS456, DYS576, DYS570, DYS481, DYF387S1, DYS627, DYS458, DYS460, DYS437, DYS439, DYS392, DYS385, DYS393, DYS391, DYS390, DYS456, DYS635, DYS449, DYS533, DYS438, DYS389I, DYS448, DYS389II, DYS19, GATA_H4, DYS518	HEALTH Gene Technologies Co. Ltd., China
43	DNATyper™15 PCR Genotyping System	D6S1043, D21S11, D7S820, CSF1PO, D2S1338, D3S1358, D13S317, D8S1179, D16S539, Penta E, D5S818, vWA, D18S51, FGA, amelogenin	Institute of Forensic Science, Ministry of Public Security, Beijing, China

(d) List of Commercially Available Capillary Electrophoresis Instruments

Sl. no.	Systems	Manufacturer	Capillary no.
1	ABI 310	Thermo Scientific	1
2	ABI 3130	Thermo Scientific	4
3	ABI 3130*xl*	Thermo Scientific	16
4	ABI 3500	Thermo Scientific	8
5	ABI 3500xL	Thermo Scientific	24

(continued)

Sl. no.	Systems	Manufacturer	Capillary no.
6	ABI 3730	Thermo Scientific	48
7	ABI 3730*xl*	Thermo Scientific	96
8	MegaBACE 500	Amersham Biosciences	48
9	MegaBACE 1000	Amersham Biosciences	96
10	MegaBACE 4000	Amersham Biosciences	384
11	SCE 2410	SpectruMedix Corporation	24
12	SCE 9610	SpectruMedix Corporation	96
13	SCE 19210	SpectruMedix Corporation	192
14	CEQ 8800	Beckman Coulter	8

(e) List of Data Analysis Software

Sl. no.	Software	Manufacturer
1	GeneMapperID and ID-X	Thermo Scientific
2	TrueAllele	(Pittsburgh, PA)
3	STRess (STR Expert System Suite) and FSS i-cubed	Forensic Science Service (FSS)
4	Open Source Independent Review and Interpretation System (OSIRIS)	National Center for Biotechnology Information (NCBI)
5	GenoProof 2 and GenoProof Mixture	Qualitype AG
6	GeneMarker	Soft Genetics, LLC (State College, PA)

Appendix C: Good Laboratory Practices

General Molecular Biology Laboratories

1. Always wear lab coats and gloves inside the laboratory.
2. Always practice a good housekeeping policy in the laboratory.
3. Gloves should be changed frequently.
4. Use paper, pens, lab coats, and other accessories dedicatedly in one lab.
5. Always maintain the unidirectional workflow.
6. Never drink or eat anything inside the laboratory.
7. Avoid mouth pipetting in the laboratory.
8. While opening/closing the centrifuge tubes, avoid touching the inner surface of the lids.
9. Never use any reagent from the stock itself; first prepare small aliquots and use from the aliquots.
10. Pulse centrifuge each reagent prior to use.
11. All the workers should have proper lab induction and training.
12. Always maintain the logbook.
13. Expose the reagents for minimal freeze-thaw cycle.
14. Reagents containing fluorophores should be stored away from light to prevent photo-leaching.
15. Always use controls during experiments.
16. Decontaminate the working area using UV irradiation, sodium hypochlorite, or 1 M HCl.

General Forensic DNA Fingerprinting Laboratories

1. The DNA fingerprinting laboratory should be designed in such a way to have a separate pre-PCR and post-PCR lab.

H. R. Dash et al., *Principles and Practices of DNA Analysis: A Laboratory Manual for Forensic DNA Typing*, Springer Protocols Handbooks, https://doi.org/10.1007/978-1-0716-0274-4,

2. The pre-PCR lab comprises evidence handling, sample preparation, DNA isolation, preparation of DNA samples for quantification, and PCR amplification.
3. The laboratory should use validated instruments.
4. The laboratory should use validated laboratory protocols (SOPs).
5. Suitable allelic ladders, positive and negative amplification controls, and reagent blanks should be used for each dataset.
6. Internal size standard should be used in each individual sample.
7. For correct interpretation, a qualified analyst or supervisor should review the result.
8. Quality assurance audits of the laboratory should be carried out from time to time.
9. All the forensic DNA analysis should be performed as per SWGDAM guidelines.
10. The laboratory should be accredited by the proper agencies to carry out DNA typing work.
11. Individual DNA analysts should undergo training, proficiency tests, and continuing education.
12. Statistical evaluation for the significance of match should be conducted for each match report.
13. Inconclusive result should be reported in situations where two analysts remain in disagreement on the data and if it is felt that insufficient information exists to draw any conclusion.
14. All the analysis should be validated for their reproducibility, robustness, and reliability.
15. The DNA laboratory must maintain documentation of all aspects of DNA analysis procedure, related documents, and interpretation of results.
16. Prior to analysis of a reagent, kit, or instrument, internal laboratory validation should be performed each time.

For Equipment, Materials, and Facilities

1. The inventory of instruments present in the laboratory should be maintained for their manufacturers, model numbers, serial numbers, agency inventory numbers, and acquisition dates.
2. Operating manual of each instrument should be available readily.
3. Calibration, maintenance procedure, and logs should be maintained for each instrument.

4. Dedicated instruments should be used for dedicated experiments.
5. Logs should be maintained for kits and other reagents having an expiration period.
6. Written validated procedures must be present for preparation of all chemicals, reagents, standards, as well as controls.
7. All the reagents should contain labels mentioning the name of the reagent, concentration, date of preparation, initial of the individual preparing the reagent, storage requirements, and its expiry date.
8. Data regarding the supplier, catalog number, lot number, receiving data, and storage location should be maintained.
9. Specific procedure should be maintained for cleaning, preparation, and sterilization in the laboratory.

For Evidence Handling

1. Each case and their exhibits should be labeled with the unique identifier that may be a number or barcode.
2. A clear, well-documented chain of custody should be maintained from time to time for all the evidences and the biological samples.
3. A written policy should be present to ensure proper storage of the exhibits in their appropriate storage conditions.
4. Care should be taken to protect the samples from loss, contamination, and degradation over time during storage in the laboratory.
5. Specific law and regulations of the agency/agencies involved should be maintained during evidence handling.

Appendix D: Allele Frequency Data for Calculation of Paternity Index

H. R. Dash et al., *Principles and Practices of DNA Analysis: A Laboratory Manual for Forensic DNA Typing*, Springer Protocols Handbooks, https://doi.org/10.1007/978-1-0716-0274-4, © Springer Science+Business Media, LLC, part of Springer Nature 2020

Table D.1
Allele frequency data and paternity and forensic parameters of the 21 genetic markers tested during the current study ($n = 551$)

Alleles	Genetic markers/loci tested: D3S1358	vWA	D16S539	CSF1PO	TPOX	D8S1179	D21S11	D18S51	D2S441	D19S433	TH01	FGA	D22S1045	D5S818	D13S317	D7S820	SE33	D10S1248	D1S1656	D12S391	D2S1338
6										0.001	0.239					0.001					
7			0.002	0.003	0.004					0.001	0.155			0.001	0.004	0.028					
8			0.073	0.004	0.339	0.007					0.142			0.002	0.232	0.219	0.012		0.040		
9		0.002	0.154	0.028	0.150	0.008			0.005	0.005	0.304			0.027	0.105	0.063	0.001		0.012		
9.3											0.141										
10		0.002	0.092	0.188	0.094	0.203		0.008	0.351	0.001	0.016		0.005	0.108	0.087	0.240			0.006		
11			0.290	0.288	0.384	0.065		0.018	0.360	0.006		0.002	0.299	0.389	0.214	0.256	0.005	0.013	0.151		
11.3									0.068												
12		0.002	0.235	0.393	0.025	0.086		0.095	0.063	0.082			0.005	0.295	0.274	0.165	0.009	0.016	0.098		
12.2										0.007											
13	0.003	0.008	0.124	0.088	0.003	0.145		0.125	0.020	0.295				0.170	0.059	0.024	0.012	0.111	0.136		0.001
13.2										0.021							0.001				
14	0.035	0.141	0.029	0.007	0.002	0.162	0.002	0.293	0.108	0.246			0.065	0.008	0.024		0.015	0.262	0.108		
14.2										0.057							0.001				
15	0.277	0.062	0.002	0.001	0.001	0.199	0.001	0.164	0.020	0.128		0.002	0.364		0.002		0.018	0.303	0.138	0.004	0.001
15.2										0.046							0.005				
15.3																	0.004		0.016		
16	0.315	0.227				0.106		0.119	0.005	0.061			0.174				0.037	0.203	0.146	0.012	0.005
16.2										0.025							0.006				
16.3																	0.002		0.039		
17	0.252	0.268				0.013	0.001	0.078		0.012			0.077				0.065	0.085	0.058	0.135	0.064
17.2										0.005							0.001		0.001		
17.3																			0.022	0.005	
18	0.107	0.205				0.001		0.035				0.005	0.010			0.001	0.083	0.007	0.009	0.246	0.192

18.2																	0.002				
18.3																			0.012	0.011	
19	0.010	0.076						0.027				0.055	0.001			0.001	0.087	0.001	0.001	0.168	0.171
19.2												0.001									
20	0.001	0.006						0.013				0.110					0.085			0.096	0.125
20.2												0.001					0.004				
20.3												0.001									
21								0.011			0.001	0.115					0.041			0.095	0.058
21.1																	0.001				
21.2												0.011					0.015				
22								0.007			0.001	0.159					0.011		0.001	0.108	0.055
22.1												0.001									
22.2												0.006					0.019				
23								0.005			0.001	0.180					0.007		0.001	0.069	0.142
23.1												0.001									
23.2												0.006					0.030				
24								0.001				0.166					0.002			0.028	0.087
24.2												0.004					0.031				
25							0.001				0.001	0.116								0.014	0.082
25.2																	0.036				
26							0.001					0.049					0.001			0.004	0.015
26.2												0.001					0.037				
27							0.008					0.005								0.002	0.001
27.2												0.001					0.051				
28							0.145					0.001									
28.2							0.001										0.054				
29							0.205														
29.2							0.006									0.001	0.059				

(continued)

Table D.1 (continued)

Alleles	Genetic markers/loci tested D3S1358	vWA	D16S539	CSF1PO	TPOX	D8S1179	D21S11	D18S51	D2S441	D19S433	TH01	FGA	D22S1045	D5S818	D13S317	D7S820	SE33	D10S1248	D1S1656	D12S391	D2S1338
30						0.001	0.173														
30.2							0.031									0.001	0.075				
30.3							0.003														
31							0.035														
31.2							0.103										0.037				
32							0.006										0.002				
32.2							0.185										0.021				
33.2						0.003	0.085										0.009				
34																	0.001				
34.2							0.008			0.001							0.005				
35																	0.001				

References

Ahmed HA, MacLeod ET, Hide G, Welburn SC, Picozzi K (2011) The best practice for preparation of samples from FTA®cards for diagnosis of blood borne infections using African trypanosomes as a model system. Parasit Vectors 4:68

Alketbi SK (2018) The affecting factors of touch DNA. J Forensic Res 9:3. https://doi.org/10.4172/2157-7145.1000424

Alshamali F, Alkhayat AQ, Budowle B, Watson N (2004) Y chromosome in forensic casework and paternity testing. Int Congr Ser 1261:353–356

Alvarez-Cubero MJ, Saiz M, Martinez-Gonzalez LJ, Alvarez JC, Eisenberg AJ, Budowle B, Lorente JA (2012) Genetic identification of missing persons: DNA analysis of human remains and compromised samples. Pathobiology 79:228–238

Bartlett T, Short S (2017) DNA rich evidence. In: Crime scene management scene specific methods, 2nd edn. Wiley, Chichester, p 145

Boattini A, Sarno S, Bini C, Pesci V, Barbieri C, Fanti SD, Quagliariello A, Pagani L, Ayub Q, Ferri G, Pettener D, Luiselli D, Pelotti S (2016) Mutation rates and discriminating power for 13 rapidly mutating Y-STRs between related and unrelated individuals. PLoS One 11: e0165678. https://doi.org/10.1371/journal.pone.0165678

Brandstätter A, Peterson CT, Irwin JA, Mpoke S, Koech DK, Parson W, Parsons TJ (2004) Mitochondrial DNA control region sequences from Nairobi (Kenya): inferring phylogenetic parameters for the establishment of a forensic database. Int J Legal Med 118:294–306

Bruijns B, Tiggelaar R, Gardeniers H (2018) Massively parallel sequencing techniques for forensics: a review. Electrophoresis 39:2642–2654

Budowle B, Allard MW, Wilson MR, Chakraborty R (2003) Forensics and mitochondrial DNA: applications, debates, and foundations. Annu Rev Genom Hum Genet 4:119–141

Budowle B, Eisenberg AJ, van Daal A (2009) Validity of low copy number typing and applications to forensic science. Croat Med J 50:207–217

Buscaino J, Barican A, Farrales L, Goldman B, Klevenberg J, Kuhn M, Lin F, Nguyen P, Salceda S, Schueren R, Smith C, Troup C, Tsou D, Vangbo M, King D (2018) Evaluation of a rapid DNA process with the RapidHIT® ID system using a specialized cartridge for extracted and quantified human DNA. Forensic Sci Int Genet 34:116–127

Butler JM (2006) Genetics and genomics of core short tandem repeat loci used in human identity testing. J Forensic Sci 51:253–265

Butler JM (2007) Short tandem repeat typing technologies used in human identity testing. BioTechniques 43:Sii–SSv

Butler JM, Buel E, Crivellente F, McCord BR (2004) Forensic DNA typing by capillary electrophoresis using the ABI Prism 310 and 3100 genetic analyzers for STR analysis. Electrophoresis 25:1397–1412

Carney C, Whitney S, Vaidyanathan J, Persick R, Noel F, Vallone PM, Romsos EL, Tan E, Grover R, Turingan RS, French JL, Selden RF (2019) Developmental validation of the ANDE™ rapid DNA system with FlexPlex™ assay for arrestee and reference buccal swab processing and database searching. Forensic Sci Int Genet 40:120–130

Côté A, Landry M, Rochette S, Gibson K, Lapointe M, Sarafian V (2008) Automated DNA extraction from large volumes. Forensic Sci Int Genet Suppl Ser 1:22–23

da Silva DA, Goes ACS, Carvalho JJ, Carvalho EF (2004) DNA typing from vaginal smear slides in suspected rape cases. Sao Paulo Med J 122:70–72

Datta P, Sood S, Rastogi P, Bhargava K, Bhargava D, Yadav M (2012) DNA profiling in forensic dentistry. J Indian Acad Forensic Med 34:156–159

Edwards A, Civitello A, Hammond HA, Caskey CT (1991) DNA-typing and genetic mapping with trimeric and tetrameric tandem repeats. Am J Hum Genet 49:746–756

Einot N, Shpitzen M, Voskoboinik L, Roth J, Feine I, Gafny R (2017) Reducing the workload: analysis of DNA profiling efficiency of case work items. Forensic Sci Pol Manag Int J 8:13–21

El-Alfy SH, Abd El-Hafez AF (2012) Paternity testing and forensic DNA typing by multiplex STR analysis using ABI PRISM 310 Genetic Analyzer. J Genet Eng Biotechnol 10:101–112

Ensenberger MG, Lenz KA, Matthies LK, Hadinoto GM, Schienman JE, Przech AJ, Morganti MW, Renstrom DT, Baker VM, Gawrys KM, Hoogendoorn M, Steffen CR, Martín P, Alonso A, Olson HR, Sprecher CJ, Storts DR (2016) Developmental validation of the PowerPlex® Fusion 6C System. Forensic Sci Int Genet 21:134–144

Fan H, Gulley ML (2001) DNA extraction from fresh or frozen tissues. In: Killeen AA (ed) Methods in molecular medicine, Molecular pathology protocols, vol 49. Humana Press Inc., Totowa, NJ

Fernández EP, Baeta M, Núñez C, Moreno SJ, de Pancorbo MM (2015) A new 17 X-STR multiplex for forensic purposes. Forensic Sci Int Genet Suppl Ser 5:e283–e285

Filho AT, Ferreira S, Oliveira SF (2013) Impact of a chromosome X STR Decaplex in deficiency paternity cases. Genet Mol Biol 36:507–510

Fung WK, Chung Y, Wong D (2002) Power of exclusion revisited: probability of excluding relatives of the true father from paternity. Int J Legal Med 116:64–67

Gefrides L, Welch K (2011) Forensic biology: serology and DNA. In: Mozayani A, Noziglia C (eds) The forensic laboratory handbook procedures and practice. Springer, LLC, Totowa, NJ

Gill P (2001) Application of low copy number DNA profiling. Croat Med J 42:229–232

Gill P, Ivanov PL, Kimpton C, Piercy R, Benson N, Tully G, Evett I, Hagelberg E, Sullivan K (1994) Identification of the remains of the Romanov family by DNA analysis. Nat Genet 6:130–135

Gopinath S, Zhong C, Nguyen V, Ge J, Lagacé RE, Short ML, Mulero JJ (2016) Developmental validation of the Yfiler(®) Plus PCR Amplification Kit: an enhanced Y-STR multiplex for casework and database applications. Forensic Sci Int Genet 24:164–175

Goray M, Eken E, Mitchell RJ, van Oorschot RA (2010) Secondary DNA transfer of biological substances under varying test conditions. Forensic Sci Int Genet 4:62–67

Grisedale KS, Murphy GM, Brown H, Wilson MR, Sinha SK (2018) Successful nuclear DNA profiling of rootless hair shafts: a novel approach. Int J Legal Med 132:107–115

Hameed IH, Jebor MA, Ommer AJ, Yoke C, Zadian HK, Al-Saadi AH, Abdulazeez MA (2014) Genetic variation and DNA markers in forensic analysis. Afr J Biotechnol 13:3122–3136

Hameed IH, Kareem MA, Rafid HH (2015) X-chromosome short tandem repeat, advantages and typing technology review. Afr J Biotechnol 14:535–541

Higgins D, Austin JJ (2013) Teeth as a source of DNA for forensic identification of human remains: a review. Sci Justice 53:433–441

Hong SL (2014) Challenges in forensic DNA profiling and critical issues in interpretation of STR profiles. J Life Sci 8:709–715

Horsman KM, Barker SLR, Ferrance JP, Forrest KA, Koen KA, Landers JP (2005) Separation of sperm and epithelial cells in a microfabricated device: potential application to forensic analysis of sexual assault evidence. Anal Chem 77:742–749

Jacewicz R, Berent J, Prosniak A, Kadlubek M, Szram S (2004) The evaluation of the IDENTIFILER system in paternity testing in Poland. Int Congr Ser 1261:538–540

Kayser M (2017) Forensic use of Y-chromosome DNA: a general overview. Hum Genet 136:621–635

Kloosterman AD, Kersbergen P (2003) Efficacy and limits of genotyping low copy number (LCN) DNA samples by multiplex PCR of STR loci. J Soc Biol 197:351–359

Krings M, Stone A, Schmitz RW, Kraintzki H, Stoneking M, Paabo S (1997) Neandertal DNA sequences and the origin of modern humans. Cell 90:19–30

Kumar A, Harish D, Singh A, Kulbhushan, Kumar GAS (2014) Unknown dead bodies: problems and solutions. J Indian Acad Forensic Med 36:76–80

Lee SB, Shewale JG (2017) DNA extraction methods in forensic analysis. In: Encyclopedia of analytical chemistry. Wiley, Chichester. https://doi.org/10.1002/9780470027318.a1104m.pub2

Lee SB, McCord B, Buel E (2014) Advances in forensic DNA quantification: a review. Electrophoresis 35:3044–3052

Lever MA, Torti A, Eickenbusch P, Michaud AB, Šantl-Temkiv T, Jørgensen BB (2015) A modular method for the extraction of DNA and RNA, and the separation of DNA pools from diverse environmental sample types. Front Microbiol 6:476. https://doi.org/10.3389/fmicb.2015.00476

Liu JY, Zhong C, Holt A, Lagace R, Harrold M, Dixon AB, Brevnov MG, Shewale JG, Hennessy LK (2012) AutoMate ExpressTM forensic DNA extraction system for the extraction of genomic DNA from biological samples. J Forensic Sci 57:1022–1030

Llamas B, Valverde G, Fehren-Schmitz L, Weyrich LS, Cooper A, Haak W (2017) From the field to the laboratory: controlling DNA contamination in human ancient DNA research in the high-throughput sequencing era. Sci Technol Archaeol Res 3:1–14

Lutz S, Weisser HJ, Meier V, Amberg R (2000) Significance of mitochondrial DNA for forensic stain analysis, identification and forensic lineage determination. Arch Kriminol 206:50–58

Macan M, Uvodiæ P, Botica V (2003) Paternity testing in case of Brother-sister incest. Croat Med J 44:347–349

Magalhães T, Dinis-Oliveira RJ, Silva B, Corte-Real F, Vieira DN (2015) Biological evidence management for DNA analysis in cases of sexual assault. Scient World J 2015:365674. https://doi.org/10.1155/2015/365674

Magnuson VL, Ally DS, Nylund SJ, Karanjawala ZE, Rayman JB, Knapp JI, Lowel AL, Ghosh S, Collins FS (1996) Substrate nucleotide determined non-templated addition of adenine by TaqDNA polymerase: implications for PCR based genotyping and cloning. BioTechniques 21:700–709

Manabe S, Morimoto C, Hamano Y, Fujimoto S, Tamaki K (2017) Development and validation of open-source software for DNA mixture interpretation based on a quantitative continuous model. PLoS One 12:e0188183

Martin P, Garcia O, Albarran C, Garcia P, Alonso A (2006) Application of mini-STR loci to severely degraded casework samples. Int Congr Ser 1288:522–525

Mulero JJ, Chang CW, Lagace RE, Wang DY, Bas JL, McMahon TP, Hennessy LK (2008) Development and validation of the AmpF*l*STR® MiniFiler™ PCR amplification kit: a MiniSTR multiplex for the analysis of degraded and/or PCR inhibited DNA. J Forensic Sci 53:838–852

Negi DS, Alam M, Bhavani SA, Nagaraju J (2006) Multistep microsatellite mutation in the maternally transmitted locus D13S317: a case of maternal allele mismatch in the child. Int J Legal Med 120:286–292

Nieuwerburgh FV, Hoofstat DV, Neste CV, Deforce D (2014) Retrospective study of the impact of miniSTRs on forensic DNA profiling of touch DNA samples. Sci Justice 54:369–372

Paabo S, Gifford JA, Wilson AC (1988) Mitochondrial DNA sequences from a 7000 year old brain. Nucleic Acids Res 16:9775–9778

Pajnič IZ, Pogorelc BG, Balažic J (2010) Molecular genetic identification of skeletal remains from the Second World War Konfin I mass grave in Slovenia. Int J Legal Med 124:307–317

Panneerchelvam S, Norazmi MN (2003) Forensic DNA profiling and database. Malaysian J Med Sci 10:20–26

Prinz M, Schiffner L, Sebestyen JA, Bajda E, Tamariz J, Shaler RC, Baum H, Caragine T (2006) Maximization of STR DNA typing success for touched objects. Int Congr Ser 1288:651–653

Raymond JJ, van Oorschot RAH, Gunn PR, Walsh SJ, Roux C (2009) Trace DNA success rates relating to volume crime offences. Forensic Sci Int Genet Suppl Ser 2:136–137

Sanchez JJ, Børsting C, Morling N (2005) Typing of Y chromosome SNPs with multiplex PCR methods. Methods Mol Biol 297:209–228

Santos GC (2018) FTA cards for preservation of nucleic acids for molecular assays. Arch Pathol Lab Med 142:308–312

Shrivastava P, Jain T, Trivedi VB (2014) Usefulness of X STR haplotype markers in forensic DNA profiling. Helix 4:582–589

Sobiah R, Syeda RH, Zunaira E, Nageen Z, Maria K, Syeda AZ, Shahana SM, Akifa M, Abdul J, Muhammad RK (2018) Implications of targeted next generation sequencing in forensic science. J Forensic Res 9:1. https://doi.org/10.4172/2157-7145.1000416

Sobrino B, Brion M, Carracedo A (2005) SNPs in forensic genetics: a review on SNP typing methodologies. Forensic Sci Int 154:181–194

Sowmyya T (2016) Touch DNA: an investigative tool in forensic science. Int J Curr Res 8:26093–26097

Tao R, Zhang J, Bian Y, Dong R, Liu X, Jin C, Zhu R, Zhang S, Li C (2018) Investigation of 12 X-STR loci in Mongolian and Eastern Han populations of China with comparison to other populations. Scient Rep 8:4287. https://doi.org/10.1038/s41598-018-22665-3

Turingan RS, Vasantgadkar S, Palombo L, Hogan C, Jiang H, Tan E, Selden RF (2016) Rapid DNA analysis for automated processing and interpretation of low DNA content samples. Investig Genet 7:2. https://doi.org/10.1186/s13323-016-0033-7

van Oorschot RAH, Phelan DG, Furlong S, Scarfo GM, Holding NL, Cummins MJ (2003) Are you collecting all the available DNA from touched objects? Int Congr Ser 1239:803–807

van Oorschot RAH, Ballantyne KN, Mitchell RJ (2010) Forensic trace DNA: a review. Investig Genet 1:14

Vandewoestyne M, Hoofstat DV, Groote SD, Thuyne NV, Haerinck S, Nieuwerburgh FV, Deforce D (2011) Source of DNA contamination and decontamination procedures in the forensic laboratory. J Forensic Res S2:001

Vaught JB, Henderson MK (2011) Biological sample collection, processing, storage and information management. IARC Scient Publ 163:23–42

Ward RH, Frazier BL, Dew-Jager K, Pääbo S (1991) Extensive mitochondrial diversity within a single Amerindian tribe. Proc Natl Acad Sci USA 88:8720–8724

Whitaker JP, Cotton EA, Gill P (2001) A comparison of the characteristics of profiles produced with the AmpFlSTR SGM Plus multiplex system for both standard and low copy number (LCN) STR DNA analysis. Forensic Sci Int 123:215–223

Wickenheiser RA (2002) Trace DNA: a review, discussion of theory and application of the transfer of trace quantities of DNA through skin contact. J Forensic Sci 47:442–450

Williams G, Pandre M, Ahmed W, Beasley E, Omelia E, World D, Yu H (2013) Evaluation of low trace DNA recovery techniques from ridged surfaces. J Forensic Res 4:1–3

Williamson AL (2012) Touch DNA: forensic collection and application to investigations. J Assoc Crime Scene Reconstr 18:1–5

Zhang S, Tian H, Wu J, Zhao S, Li C (2013) A new multiplex assay of 17 autosomal STRs and amelogenin for forensic application. PLoS One 8: e57471

Further Reading

Butler JM (2012) Advanced topics in forensic DNA typing: methodology. Academic Press, Amsterdam. ISBN: 978-0-12-374513-2

Dash HR, Shrovastava P, Mohapatra BK, Das S (2018) DNA fingerprinting: advancements and future endeavors. Springer, Singapore. ISBN: 978-981-13-1582-4

Jamieson A, Bader S (2016) A guide to forensic DNA profiling. Wiley, Boston, MA. ISBN: 978-1-118-75152-7

Taupin JM (2013) Introduction to forensic DNA evidence for criminal justice professionals. CRC Press, Hoboken, NJ. ISBN: 978-1-439-89909-0

MIX
Papier aus verantwortungsvollen Quellen
Paper from responsible sources
FSC® C105338

If you have any concerns about our products,
you can contact us on
ProductSafety@springernature.com

In case Publisher is established outside the EU,
the EU authorized representative is:
Springer Nature Customer Service Center GmbH
Europaplatz 3, 69115 Heidelberg, Germany

Printed by Libri Plureos GmbH
in Hamburg, Germany